VOLUME 95

THE ORIGIN OF CANCER

TAUTOMERISM & METHYLATION

EDITION 1

Carlos L Partidas

DEDICATION

THE ELECTRONIC BODY OF EVERY HUMAN BEING, WHICH WE CAN
AFFECT CHEMICALLY TO CAUSE OURSELVES THE FIVE ORGANIC
DISEASES, DUE TO THE WAY WE FEED OURSELVES WITH THE MEAT OF A
BROTHER ANIMAL

CONTENTS

ACKNOWLEDGEMENT

THE HUMAN BEING WHO WISHES TO ACQUIRE THE KNOWLEDGE OF HOW HIS
PHYSICAL BODY WORKS CHEMICALLY

Chapter 1
THE ORIGIN OF THE UNIVERSE

Once we are certain of how the Universe originated, we are obliged to rewrite the chemistry of disease; for the organic diseases that occur in the electronic body are the result of the mismatched chromosomes in the nucleus of cells at the microscopic level. In this case, we refer to the five organic diseases, among which is cancer. Cancer is caused by an error in the electronic coupling between the bases adenine-thymine and guanine-ketone-cytosine. These are organic diseases, because we ourselves induce these mismatches between the bases of our DNA by the way we feed ourselves with cells of animal origin.

These electronic mismatches between the chromosome-forming bases in the DNA are due to the incorrect coupling of electronic charges, which are generally known as Van del Waals forces, i.e., hydrogen bridges. If we un-consciously manage to alter these couplings between the bases of our DNA, we will be causing a mutation to the chromosomes in the cell nucleus, since our chromosomes will be forced to electronically couple the bases in a dif-ferent order than the one that was originally established.

The electronic effect by which we force cells to make these electronic changes is tautomerism and methyla-tion. As can be seen in Figure 8 below, through tautomer-ism, the guanine base is induced to convert from its nor-mal ketone form to an alcohol, which creates an imbal-ance in our DNA known as keto-enolic balance.

The consumption of a different kind of cell than our own, i.e. cells of animal origin, induces a keto-enolic bal-ance in our cell nucleus, because the consumption of an-imal meat cells will increase the acidity of our body fluid, and if there is an increase in acidity in the body, there will be an increase in acidity in the nucleus of our cells, and our ketonic guanine base will be converted to an enolic base.

When our cells cease to function, we do not discard them whole; rather, from the extinct cells, our catabolic system will convert our adenine and guanine purine bases into sodium urate. Sodium urate is our main antioxidant; for it is sodium urate that neutralises the free radicals that remain after haemolysis. In other words, sodium urate neutralises the free radicals that remain after the red blood cells have ceased to function as transporters of oxygen and carbonic acid in the blood. The antioxidant sodium urate also prevents early oxidation of our healthy red blood cells.

Cells of animal origin are chemically the same as ours, only the chromosomes of an animal or the chromosomes of a human being insert the bases in a different order, which forms a genetic code that makes us look physically different from an animal. The genetic code makes all living things look physically different, even though all living things have chemically the same kinds of cells. In other words, all living beings are consubstantial, or all living beings are physically made up of the same kinds of cells.

Thus, the consumption of meat provides us with the same adenine and guanine purine bases that are in the DNA of the dead cells of the animal's meat. Through our catabolism, the adenine and guanine purine bases in the DNA of the ingested meat are converted in parallel into sodium urate.

This shows us that the human being is a mutation of another living being, but the mutated living beings separated into castes when the intermediate mutations failed to reproduce. It seems that we are talking about the present time, but these mutations have been going on continuously for billions of years, which, comparatively speaking, is an instant in comparison to the existence of human beings on earth.

If our blood is not acidic, the breakdown of our proteins does not generate the methyl group from our amino acid methionine. However, when we consume animal protein, we cause methylation, because the amino acid methionine in the animal protein will leave the methyl group free, which contributes to a genetic error, i.e., caused by the way we eat.

The amino acid methionine is contained in all animal proteins because methionine is the initiating amino acid for protein synthesis by ribosomes. Methylation is produced by the methyl group of the amino acid methionine that comes with animal protein; and, when the amino acid methionine loses the methyl group, the amino acid methionine is converted into the amino acid homocysteine.

The amino acid methionine is carried by all proteins of animal origin, as methionine is the first amino acid carried by messenger RNA leaving the nucleus of cells into the cytoplasm in triplet form. Methionine is the amino acid that signals the ribosome to 'start here'; that is, methionine is the initiating amino acid for the ribosome to begin synthesis of a given protein. Then, the ribosome will continue to insert in an order the next amino acids brought in by the messenger RNA. Each amino acid is carried by the transfer RNA; that is, each transfer RNA carries only one amino acid. In the order indicated by the messenger RNA, the ribosome builds the protein chain. Then, when all the amino acids of the messenger RNA have been inserted, the messenger RNA brings a triplet at the end of the chain that tells the ribosome: 'nothing goes here'; so, when it reaches the triplet brought by the messenger RNA, the ribosome completes the synthesis of the protein chain.

The protein that is built up will replenish the bone system and replace the proteins that have ceased to function, such as the proteins that line organs, nails, hair, cavities, arteries, veins, etc.

When the meat of an animal is consumed, the meat cells of the animal are consumed. So, from the DNA of the ingested meat cells, sodium urate will be produced, but this exogenous sodium urate will be in excess of our normal sodium urate; therefore, our sodium urate ↔ uric acid balance will be shifted to a higher value; and, because of this imbalance, we will have excess uric acid. Excess uric acid produces an increase in body acidity, or a decrease in the pH of our blood; so, the amino acid methionine ingested from animal protein will lose the methyl group, and the methyl group detached from the amino acid methionine will convert the cytosine base to the thymine base.

Likewise, because of the increase in body acidity, the uracil base will undergo tautomerism; and, after tautomerism has occurred to the uracil base, methylation will occur to this uracil base. Methylation will also convert the uracil base to the thymine base.

Due to tautomerism and methylation, the nucleus of the cell will be left without the cytosine and uracil bases, and the chromosomes will form the wrong base-pair couplings: adenine-thymine and guanine-enolic-thymine.

As can be seen in Figure 7, the adenine base has 2 ketone groups, but the thymine base cannot be tautomerised, as the thymine base has its aromatic ring complete with its double bonds. We can see in Figure 7 that the adenine base has no ketone groups, so tautomerism cannot happen to the adenine base.

This erroneous DNA is faster to synthesise, so the reproduction of the mutated cells is accelerated compared to the cells that remain healthy; and, because of this mismatch, cancer is produced due to the speed with which the mutant cells replicate compared to the healthy cells. The mutated cells are normal cells, but these mutant cells reproduce faster than healthy cells, resulting in a lump of amorphous cells in the form of a tumour, which is clinically described as cancer.

In Figure 1, we can see the first 3 energy levels N-1, N-2 and N-3 that formed the Universe. In each energetic level there are 5 almatrinos; since, the energy of the Universe is produced by energetic levels conformed each one by 5 almatrinos. In Figure 1 we can see the first 15 almatrinos that formed the Universe.

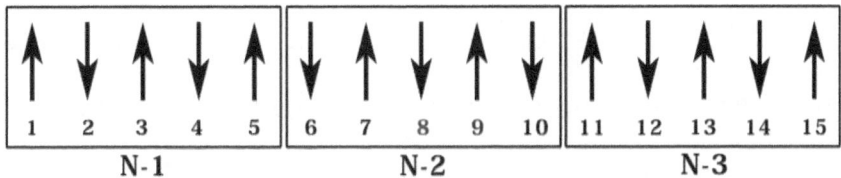

FIGURE 1
FORMATION OF THE 3 FIRST LEVELS N-1, N-2 AND N-3 THAT CREATED THE UNIVERSE

An almatrino is the minimum amount of electronic energy that was formed in nothingness; and this minimum amount of electronic energy, being energy, began to rotate towards nothingness; therefore, electronic energy rotates and still rotates against nothingness; that is to say, in nothingness there is nothing that can stop the accelerating movement of the electronic energy of the Universe. Therefore, the Universe was not created; rather, the Universe is still creating itself, because in nothingness there is nothing that can stop the expansive motion of the electronic energy of the Universe.

As can be seen in Figure 1, the electronic energy of the Universe is generated in an alternating manner; whereby, the first two electronic energies 1 and 2 repel each other, while the two energies 1 and 3 attract each other. These forces of attraction and repulsion are what causes the cohesive force of gravity to form, which is the energy that forms the physical space of the Universe.

The electronic energy is still spinning against nothingness; therefore, 13.8 billion years after the initial moment of an almatrino, the movement of the Universe has not stopped, nor can it stop as long as the movement of the electronic energy exists; since, when there is more movement of the electronic energy, more electronic energy is produced. That is to say, the movement of the Universe will not stop; the movement of the Universe is an accelerated movement; so, the movement of the Universe cannot stop at a point, even if that point is located in the most infinite. The Universe is and will be forever eternal. The minimum amount of electronic energy of an almatrino could not be stopped; therefore, the expansion of the Universe is exponential of the form $y=e^{xt}$.

However, the spin speed of the almatrino could not have an infinite value; therefore, there is a limit to the speed, where the electronic energy is converted into electronic matter. The energy equation that determines at what spin speed the electronic energy is converted into electronic matter is the energy equation of Albert Einstein and Mileva Marić, $E=mC^2$. However, Albert Einstein and Mileva Marić considered the initial mass of the Universe to be imaginary; that is, for Albert Einstein and Mileva Marić the initial mass of the Universe does not exist. However, if the electronic energy of the Universe is real, and the electronic matter of the Universe was formed from the integration of electronic energy, then the initial electronic matter of the Universe is likewise a real quantity. Furthermore, we can see this electronic energy condensed in the form of electronic matter forming the stars, suns, planets, or the electronic matter forming the physical bodies of all living things that exist on Earth. Albert Einstein and Mileva Marić confused magnetic mass with electronic matter.

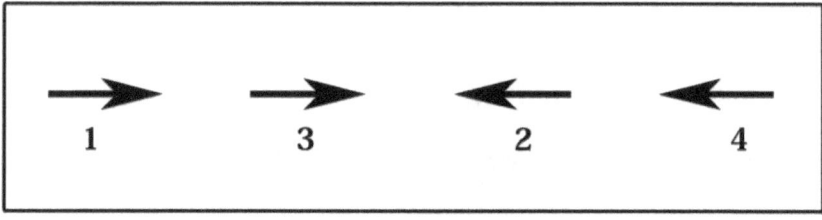

N-1

FIGURE 2
THE MAGNETIC ENERGY ROTATES FROM RIGHT TO LEFT AND FROM LEFT TO RIGHT AT THE FIRST N-1 ENERGY LEVEL OF THE UNIVERSE

In Figure 2, we can see that the magnetic energy is produced by the flow of the electronic energy in Figure 1. In Figure 1, the electronic energy of the upward flowing almatrinos 1 and 3, produce a magnetic energy 1 and 3 that rotates from right to left, as can be seen in Figure 2. Magnetic energies 1 and 3 integrate and form the positive magnetic mass of a male being.

As can be seen in Figure 1, the two downward flowing electronic energies form the 2 magnetic energies 2 and 4 rotating from right to left, as can be seen in Figure 2. The two negative magnetic energies 2 and 4 are integrated, and the negative magnetic mass of a female being is formed.

In both cases, male and female refers to a mammal or oviparous of any living being on Earth. The energy that provides the force for the integration of the two magnetic energies is the energetic force of memory. The energetic force of memory is what enables the energetic being to have self-awareness and to evolve in both the physical world and the spiritual world. Magnetic memory is what allows us to move forward or backward in time. It is magnetic memory that gives us the ability to orient ourselves in physical space. Magnetic memory is what gives each spirit its identity. Magnetic memory is the integrating energy of the magnetic mass; therefore, magnetic memory cannot be separated from spirit. A spirit without its magnetic memory would be the same as electronic matter; that is to say, a spirit without magnetic memory would have no knowledge of itself.

The difficult thing would be to know on what energetic level the magnetic mass that forms the silhouette of every living being on Earth was formed. Because the magnetic mass of the spirits cannot be seen by all of us from our physical electronic world.

So, neither Albert Einstein nor Mileva Marić managed to see that the magnetic mass of the spirits is different from the electronic matter of the body. So, we had to solve the imaginary value of the electronic matter of the equation of Albert Einstein and Mileva Marić, arriving at the energy equation of the Universe, which is of a real form: $Ev=m_0C^3$. In this energy equation, m_0 is the initial amount of matter in the Universe. We can project this equation in time; so, by means of the equation $Ev=m_0C^3$, we can know, for example, how much electronic matter the Universe will have at a point that is close to infinity or how much electronic matter the Universe will have in 27.6 billion years; that is, twice the age of the Universe at this moment.

In the same way, the flow of electronic energy allows us to know how the nuclei and electrons that make up electronic matter were formed. From Figure 1, the positive 1 and 3 almatrinos, or those with an upward flowing electronic energy, spontaneously integrate and the electronic matter of a positive electronic nucleus is formed. From the same Figure 1, the electronic energy of the 2 and 4 almatrinos, or those with downward-flowing electronic energy, spontaneously coalesce and the negative electronic matter of an electron is formed. The negative electrons revolve around the positive nuclei because of the differences in electronic charges and thus the atoms of the electronic matter are formed. The atoms form the molecules, and the molecules form the electronic matter where we include the electronic body of all living things on Earth. The silhouette of the magnetic mass of a man and a woman cannot be seen by everyone, and we will not be able to see them, because spirits, although they are integrated from magnetic energy as magnetic mass, the magnetic mass of the spirit does not contain electronic matter.

After the mating of the two physical bodies, male and female, a spirit from the spirit world will be incorporated into its new physical body. In the human race, the coupling will take place 5 months after gestation; therefore, at 9 months, the living being of a boy or a girl will be born, depending on the sex of the physical body that has been formed in the pregnant womb; which will depend on the degree of acidity of the vagina and the blood of the mother of the baby at the initial moment of gestation.

Outside the Universe there is nothing, nobody could have created the Universe; since, in order to create the Universe, that somebody must have been in nothingness. But, in nothingness there is nothing. The nothingness is not infinite, because in the nothingness there is nothing; the nothingness is an imaginary line that we use to delimit the external border of the Universe, to be able to say that the Universe expands in an exponential way towards where nothing exists. In other words, the Universe is imploding into an absolute void, because outside the Universe there is nothing.

The cells that make up the physical body of any living being are made up of electronic matter that is changeable as time goes on, because cells arose from the mutations of viruses. The perfection and coordinated logistics that we can observe in terms of the functionality of the body of a living being is due to the differences in the electronic charges that are microscopically happening and are still happening in the electronic bodies of living beings.

Time only exists on Earth because the matter of the electronic body changes and ages. Whereas, for the magnetic mass of the spirit there is no time, because the magnetic mass of the spirit is not changeable. It is not changeable, because the magnetic mass of the spirit contains no electronic matter at all. The magnetic mass of the spirit is eternal; and, being energy, the magnetic mass of the spirit can live anywhere in the Universe. The only ones that can live as spirits and bodies are the living beings that exist on planet Earth. As said, the magnetic mass of the spirit does not contain electronic matter; therefore, the magnetic mass of the spirit is not affected by the energetic force of gravity nor by low or very high temperatures.

A virus is a unit that forms the smallest 'life' that can exist in the Universe, because viruses are the micro-bodies that formed in the first energetic levels of the Universe. It will be difficult for us to know at what energetic level viruses were formed but these micro bodies of viruses were formed at the first energetic levels of the Universe before the spirits of the higher animals; and, among which, we can include the spirits of human beings. Viruses are formed by the integration of microelectronic matter with the micro magnetic mass that is spread throughout the Universe. Viruses are microelectronic bodies that live dormant, they just wait for the right conditions to be able to replicate spatially.

The electronic bodies of the Universe were very hot, because in the beginning, the size of the Universe was infinitesimal; so, at that micro size, the micro-Universe could not withstand the enormous amount of energy that is produced by the flow of electronic energy and the spinning of the magnetic energy that produces the flow of electronic energy. Millions of years passed; and the Earth already formed at a very high temperature began to cool; and, as the Earth cooled, conditions were achieved on Earth for the viruses to awaken from their dormant state. Thus, on Earth, viruses mutated into the form of prokaryotic and eukaryotic cells. These prokaryotic and eukaryotic cells mutated further, thus giving rise to the physical form of all living things on Earth, i.e., from an insect to a human being.

The cells of living beings continued to mutate, and the different races of animals were formed, as well as the different castes of human beings. In such a way that all living beings are a mutation of the cells, which we will be able to modify electronically, since the cells are adaptations that are happening electronically, due to the differences generated by the different electronic charges that produce the logistics that we classify as perfect; but the human being modifies this logistics of the living body with the way he feeds himself with the flesh of his brother animal.

For example, in order to be able to move the physical body, electronic energy must be supplied to the physical body. This electronic energy is produced in the cells in the form of heat to keep the cells and the physical body in motion; whereas, to move the spirit we do not need electronic energy; for, the spirit contains no electronic matter. The spirit is only the conductor that guides the electronic matter of the physical body and feeds it. The spirit is made up of magnetic mass; so, for the spirit, the electronic matter of the body is as if it does not exist. In reality, the spirit can pass through any physical body that is made up of electronic matter, because the magnetic mass of the spirit does not interact with the electronic matter. Therefore, the spirit can separate from the electronic matter that forms its physical body when the matter of the physical body no longer functions to house it as a spirit. Therefore, the death of the spirit does not exist, what exists is a separation of the magnetic mass of the spirit from its physical body formed by electronic matter.

So, we need to know how the cells of the physical body function chemically, to see how we can affect the electronic couplings between the bases that make up the DNA, to cause the physical body to undergo the different kinds of cellular mutations, which is what causes us to suffer from the five organic diseases. The diseases that we cause ourselves by the way we feed ourselves are five; and we cause these five diseases by the way we feed ourselves with the flesh of a brother animal. These five diseases are: cancer, diabetes, arthritis, heart attack and memory loss, i.e., Alzheimer's disease.

The weakening of the insulator that covers the neurons so that they do not come into contact with each other is myelin. The lack of myelin insulation causes a short circuit to occur between the neurons in the physical body, which is known as Parkinson's disease and the hereditary neurological disease Huntington's disease.

In this book, we are going to refer specifically to cancer, because if we were to consider all organic diseases in one volume, the analysis would be very extensive. The idea is to make us all understand the chemistry of the electronic body and the origin of the organic diseases that can chemically affect our physical body. As spirits we are obliged to take care of our physical body.

Chapter 2
CELLULAR RESPIRATION

Healthy cells produce heat energy in the mitochondria from the combustion of the sugar glucose with oxygen. In the final process of enthalpy ΔH, carbon dioxide is generated as a waste product. When oxygen does not reach the cells via normal respiration, the mitochondria will resort to the process of glucose fermentation; this is known as glucose glycolysis. If caloric energy production goes via glycolysis, lactate will be generated in the mitochondria instead of carbon dioxide. When cells are healthy, the mitochondria use these two pathways to produce caloric energy; since, the process for energy generation will depend on the form of respiration of the living organism. For example, when we were a spermatic cell, there was no oxygen in the mitochondria that were in the tails of the sperm; so, at that time, the sugar for the production of caloric energy was not the sugar glucose but the sugar fructose. As there is no oxygen in the sperm because the spermatic cell has no blood, in the process of enthalpy or heat generation with fructose sugar, no lactate is generated but glucose and galactose. The glucose sugar will be used by the spermatozoon for further movement as a zygote, while the galactose will be for the growth of the nervous system in the next stage of the zygote's growth i.e., for the development of the foetus. Thus, the sugar fructose is most abundant in the semen of the mammalian testes.

It will be fructose, because the heat that keeps the sperm alive is generated in the sperm cell; and it will be fructose for the initial movement, because the fructose sugar will provide the sperm with the energy necessary for the sperm to reach the egg, after the semen is leached of its alkalinity by the acidic fluid of the vagina at the moment of ovulation. When the sperm enters the egg, the sperm replicates, and the composite cells of the zygote will form, and it is from the zygote that the embryo will form. The cells of the embryo consume more energy to replicate, so when the first blood vessels appear in the embryo, the heat energy and movement for replication is provided to the foetus by glucose and oxygen from the mother's blood.

Thus, the supply of nutrients to the embryo will depend on the correct or incorrect respiration of the mother, because if not enough oxygen reaches the mitochondria of the embryo's cells, the embryo's mitochondria will resort to the process of glucose fermentation, but not through the fermentation of fructose. Therefore, lactate will be generated inside the embryo's cells.

When the mother resorts to rest through sleep, the form of respiration will be deeper; therefore, the amount of oxygen will be higher, and lactate will be converted back to pyruvate and pyruvate will be converted back to glucose. So, the life of the zygote, when the zygote transforms into an embryo, then into a foetus and into a baby, the breathing of the baby's cells will be through that normal pathway of glucose with oxygen; but, this process of the baby's normal breathing will depend on the normal breathing process of the baby's mother, because the ba-by's breathing will depend on the mother's way of breath-ing.

Then, we will come to the moment of the birth of the baby; but, the newborn baby does not have teeth nor does it generate enough saliva in her or his mouth; so, the baby does not have the enzyme amylase; but, it will not be able to eat fruits and starches to obtain from fruits the sugar fructose and it does not have in its mouth the en-zyme amylase to obtain glucose from starch. However, nature has placed the sugar lactose in its mother's milk. It is lactose, because from lactose the baby can get the sug-ars glucose and galactose. With glucose, the baby can get the caloric energy in the mitochondria of its cells for the movement of its muscles. While from the sugar galactose, the baby can get the basic nutrients for growth and con-solidation of the nervous system.

However, to obtain from these compounds that are necessary for the strengthening of its nervous system from the sugar galactose, the baby, instead of the enzyme amylase in its mouth, has the enzyme lactase in its small intestine. The enzyme lactase is the enzyme that allows the baby to break down the fats in the mother's milk to obtain fatty acids from these fats. Fatty acids have a hormone-like function in the body. The enzyme lactase is similar to the enzyme renin, which newborn calves have in their fourth stomach. In the old days, calves were slaughtered to remove the fourth stomach, where the renin is located, to produce cheese from cow's milk. Until the cow's milk powder industry came along and the calf was allowed to live longer so that the calf's mother could produce more milk, and so that the dairy industry could produce more milk powder.

Let's say, this is the normal process for the journey that a mammalian being has to go through from sperm to zygote, to embryo, to foetus, to baby in the womb and finally to baby after birth. This respiratory process for a mammal involves the fermentation of the sugar fructose, until there was an exchange of the sugar glucose with oxygen; because, when we were a sperm cell we had no blood, so that oxygen could get to us, and heat was produced by oxygen with the sugar glucose. Also, because the glucose would ferment in the sperm cell, and lactic acid would be produced inside the sperm cell.

Thus, the anaerobic pathway of respiration in the spermatozoon, although the fastest way to obtain energy, is not an efficient process, but was initially sufficient, due to the small size of the cell that forms the spermatozoon.

The sperm is immersed in a viscous fluid in order to reduce its mobility, i.e., so that the sperm is not energetically depleted by the movement.

The lower pH, i.e., the higher degree of acidity in the vagina at the time of ovulation, has the purpose of dissolving the alkaline mantle in which the spermatozoa is enveloped. The high degree of acidity in the vagina causes the alkaline mantle to dissolve, so the sperm is released and quickly reaches the egg. When the sperm reaches the egg, a change of the kind of sugar for respiration must take place; the change will be from fructose to glucose; so that, instead of fructolysis of fructose, carbon dioxide will be produced as a waste product after the reaction of oxygen with glucose. In this way, energy in the form of heat will be generated more efficiently in the mitochondria, according to the demand of the numerous cells that will be formed in the embryo.

This change of the sugar class is necessary, as a higher yield will be needed for the production of energy for the movement of the chromosomes that will produce the muscles that will grow the physical body of the embryo from the DNA replication.

As for the formation of DNA in the embryo, the combination of a sugar with the bases guanine, adenine, cytosine, thymine, and uracil will form a nucleoside. For example, the sugar deoxyribose attached to adenine forms adenosine, while esterification of the same sugar with one of these bases plus phosphoric acid will become a nucleotide. Two of the nucleotides necessary for oxygen storage and subsequent caloric energy production in the mitochondria of cells are: adenosine diphosphate ADP, and adenosine triphosphate ATP; which, although they do not play a role in the formation of the order of insertion of the bases in DNA; that is, of the genetic code that will form the physical body, these nucleosides and nucleotides, are necessary for the generation of the caloric energy that is produced in the mitochondria of the cells, thanks to the metabolism of glucose with oxygen.

Mitochondria are considered to be prokaryotic cells because mitochondria have no cell nucleus; but mitochondria produce their own DNA called mitochondrial DNA.

Initially, through glycolysis, two molecules of ATP are obtained for each molecule of glucose used, compared to the 36 molecules of ATP that are generated by the route of complete combustion of glucose to carbon dioxide, when glucose is used for normal respiration with oxygen in the mitochondria of cells. But in muscle cells, the glycolysis of anaerobic respiration i.e., without oxygen, cannot be sustained indefinitely, as lactic acid is generated from lactate in this anaerobic process. But, to prevent lactic acid from being formed, the antioxidant enzyme system was formed within the embryo's cells; therefore, lactate is not converted to lactic acid; instead, lactate will be oxidised to pyruvate.

When pyruvate is formed in the liver, either this pyruvate will be oxidised to carbon dioxide, or it can be converted back to glucose to maintain the energy reserve, mainly by resting after sleep. At the same time, the pathway back to normal aerobic combustion of glucose with oxygen can be resumed. Thus begins the cycle of cellular respiration in the embryo, when the oxygen supply can arrive via the bloodstream from the mother's blood.

After birth, the supply of ATP will be achieved by the normal route of the baby's respiration, i.e., it will be more aerobic than anaerobic, because both processes are needed, i.e., with oxygen or without oxygen, because the process of respiration will depend on the circumstances that arise in the mitochondria of the cells. For example, more muscle cells will be formed; therefore, there will be a greater demand for electronic energy for the movement of the muscles that make the baby's physical body grow, which is why the baby sleeps longer than an adult, and an elderly person may sleep less than an adult. More energy output is needed to obtain the caloric energy, such as the 36 molecules of ATP. To prevent lactic acid from forming inside the cells, the glycolytic enzymes glutathioneSH, vitamin C, and the enzymes superoxide dismutase and catalase are needed inside the cells.

If we are not aware of how this process of respiration for obtaining heat energy by the mitochondria of the cells is, we can damage the chemical structure of the cells, since the mitochondria of healthy cells need oxygen and glucose for the generation of electronic energy in the form of heat, in order to maintain cellular activity, i.e., to be able to live in a normal way.

But, if no oxygen reaches the mitochondria, to obtain heat energy, the mitochondria will resort to the process of glucose glycolysis.

That is, low oxygenation and high acidity is favourable for mutant or cancer cells, because when the body acidity is high, or the pH is low, haemoglobin will be bound to carbonic acid, which will disable haemoglobin, so that haemoglobin will not transport oxygen efficiently to the periphery of the cells. For this reason, the blood pH must change in the lungs so that haemoglobin can transport oxygen from the lungs to the periphery of the cells to exchange it for the carbonic acid taken out of the cells by myoglobin.

Myoglobin is a smaller molecule than the haemoglobin molecule, because myoglobin has only one heme group, whereas haemoglobin has four heme groups, so each haemoglobin molecule can carry four oxygen atoms.

Inside the cells, the mitochondria will produce heat energy from oxygen with glucose, leaving carbon dioxide as a waste product of combustion, which will be converted into carbonic acid by the enzyme carbonic anhydrase. This means that inside the cells, the acidity value is relatively higher, so that the myoglobin can draw the carbonic acid to the outside of the cells. On the outside of the cell, the acidity value is higher, so that haemoglobin releases oxygen and binds with carbonic acid to carry it to the lungs when haemoglobin is swept into the bloodstream.

In the alveoli, the acidity value is lower, so the carbonic acid is broken down into water vapour and carbon dioxide, and as the haemoglobin is released, it rejoins with four more oxygen atoms. The haemoglobin is then swept back into the bloodstream towards the periphery of the cells.

This process of acidity change is both chemically interesting and spectacular, but it is also tricky, because we call a pH value of neutral or 7.00 acid inside the cells, whereas in the lungs the acidity value is 7.40. Although we are talking about logarithmic quantities, this range of acidity is very narrow; therefore, we should not damage this magnificent process, so as not to affect the respiration of our cells that make up the matter of the physical body where the magnetic mass of the spirit dwells.

If the blood is more acidic than normal, the haemoglobin will not be able to get rid of the carbonic acid; therefore, the transport of oxygen to the mitochondria of the cells will not take place; and if there is no oxygen in the mitochondria, they will produce energy by the second way; that is, by the fermentation of glucose or glycolysis.

There needs to be a chemical balance within the cells. For example, a reducing enzyme system is needed so that NAD can reduce the iron III in haemoglobin back to iron II, but it is the same NAD that oxidises the iron II in haemoglobin to iron III, so that myoglobin can take the carbonic acid out of the cells as iron III and carry the oxygen into the inner part of the cells when the iron in myoglobin is as iron II.

On the outside of the cell, as mentioned, the acidity is higher, so the myoglobin gets rid of the carbonic acid that the haemoglobin captures, and the myoglobin captures the oxygen that the haemoglobin has. This is a logical exchange of complex ions by a change of acidity, but this is not a chemical reaction.

With higher acidity at the periphery of the cell, haemoglobin is left as iron III in the iron of the heme group; so haemoglobin captures carbonic acid; and, the bloodstream carries haemoglobin back to the lungs; but, in the lungs the acidity is lower; so the iron in the haemoglobin changes back from iron III to iron II, so that the haemoglobin releases the carbonic acid and re-bonds each heme group as iron II to an oxygen atom in the air entering the alveoli. Again, the bloodstream carries the oxygen-loaded haemoglobin into the cells. It's a back and forth, a giving and receiving that these two molecules do. It is the most interesting and spectacular chemical process in the Universe; for it is these two molecules that maintain the respiratory activity that sustains life in the physical world.

Let's say, this is the process for the normal respiration that takes place inside and outside the cells. But this process of exchanging carbonic acid for oxygen and oxygen for carbonic acid does not correspond to a chemical reaction, which is why Dr. Max Ferdinand Perutz called it the cooperative effect. It is to Dr. Max Ferdinand Perutz that we owe this description of the amazing respiratory process of cells. Although Dr. Perutz based his description on the measurement of the oxygen partial pressure value of 100 mmHg in the lungs and 40 mmHg in the muscle, as these would be the variables that Dr. Ferdinand Perutz could measure. But we deduce that the change of these values is rather due to a change of acidity in the cells since the process of respiration is not due to a change of oxygen gas pressure but to a change of acidity.

The higher acidity value at the periphery of the cell is called the Bohr effect, and we owe the description of this process to the Danish physicist Niels Henrik David Bohr, whose research also influenced today's understanding of the atom and quantum mechanics. Niels Bohr, like the

Belgian mathematician, astronomer, and physicist Georges Henry Joseph Édouard Lemaître, was a theologian, so Niels Bohr had his debates with Albert Einstein, who was critical of the physical, philosophical, and theological interpretations that Niels Bohr derived from physical principles and certain aspects inherent in the new quantum mechanics. For Niels Bohr's theories assumed that ordered matter is derived from a previous state, in which matter is in permanent disorder. But the disorder of electronic matter arises from an order, and it is what we know today as chaos or entropy.

It was a philosophical statement by Democritus, who created the term atom to refer to the smallest amount of physical matter in the Universe. Although today we know that elementary particles exist; and the smallest particle that exists in the Universe is an almatrino, and the smallest form of physical life that exists in the Universe is a virus.

But the subsequent experimental analysis of Dr. Ferdinand Perutz is based on the observation of the German scientist Otto Heinrich Warburg that cancer cells reproduce in an acidic medium and in an oxygen-free environment.

Thus, after birth, this narrow range of acidity for the process of respiration inside and outside the cells can be affected by nutrition, mainly due to a lack of knowledge of the chemistry of this process. If prior to gestation our parents do not know how to feed the zygote, after birth, we will carry the tradition or the force of this nutritional error by means of a custom, and we will be able to influence with our guilt, so that the antioxidant enzymes inside the cells are affected biochemically; specifically inside the cell nucleus; and thus, we will be able to change the form of the couplings between the bases by the effect of tautomerism and methylation.

The change in base coupling in DNA is what is known as a mutation, which will lead us to the origin of cancer. It is a mutation because, through the effect of tautomerism and methylation of electronic matter, the coupling of the bases that make up the chromosomes in DNA must necessarily adapt to the new changes in electronic config-uration that we cause them to undergo.

So, the correct or incorrect coupling of these adenine-thymine and guanine ketonic-cytosine bases depends on the original chemistry within the nucleus and chromo-somes of the cells. So, the chemistry of our cells will ulti-mately depend on us; for it is we as the mass of the spirit that drives the electronic matter of the body who decide how we feed ourselves; for the act of feeding when we grow up is a voluntary act.

This is a clear example of how we can influence these electronic mismatches, since the nature of electronic en-ergy is changeable due to differences in electronic charges, so that these DNA bases can mutate through a great diversity of situations and forms, or according to the chemical and physical situations in the physical world. Whereas the magnetic mass of the spirit will remain un-changed, since the way the spirit acts depends solely on the knowledge of how the electronic matter of the phys-ical body functions chemically; that is to say, it is knowledge that will awaken the consciousness of the human being. The purpose of the knowledge that this book is intended to convey is that the magnetic mass that forms the silhouette of the spirit can correctly direct the action of the electronic matter of the physical body.

The connection, i.e., the anchoring of the magnetic mass of the spirit with the electronic matter of the physical body will depend precisely on the chromosomes in the nucleus of the cells. However, the good or bad functionality of the electronic matter of the physical body, or according to the kind of use we want to make of it, will not depend on the electronic matter of the body, but on the knowledge of the magnetic mass of the spirit, which guides the electronic matter of the physical body. It is knowledge that frees us, and it is knowledge that gives the degree of consciousness to the magnetic mass of the spirit; that is, of us, who are actually, with our acquired state of consciousness, the directors of the physical body in a temporary way.

In reality, that humanity is on the verge of a transition through this learning process, so that a change of consciousness can then take place in humanity, without the need for humanity to destroy itself, and destroy the life form that exists on planet Earth.

The main cause of the high acidity in the blood and within the cells is when human beings feed on the flesh of a being whose cells are chemically identical to those of a human being, for only the position of the bases in the DNA, i.e., the genetic code, changes. By eating the flesh of a sibling, mankind is sinking into a genetic abyss of its own, mainly due to a lack of knowledge of how its electronic body functions. Through this behaviour, which is also of energetic origin, a great number of human beings suffer from organic diseases, but equally, many animals suffer from the evil behaviour of unconscious human beings, who mistreat and kill animals in order to devour them. Animals are our brothers and sisters, because all beings arise from the electronic and magnetic energies emanating from the Universe.

A low-oxygen environment in healthy cells must have started as soon as uric acid in its enolic form became more strongly bound to haemoglobin. So, this enolic uric acid did not allow haemoglobin to transport oxygen, as haemoglobin became permanently bound with the enolic uric acid. This will have to happen in this way until there is a change of acidity in the blood of the human being, i.e., for the haemoglobin to free itself from the enolic uric acid and re-bind with the oxygen in the lungs. So, that haemoglobin can transport oxygen from the lungs to the cell wall and from there to make oxygen available to the mitochondria of healthy cells. This is the only way to get the healthy cells to return to aerobic energy production via the oxidation of glucose with oxygen; by changing the acidity of the blood, uric acid from its enolic form back to its normal ketonic form. Uric acid in the ketonic form is weaker than uric acid in the enolic form, so that uric acid in the enolic form binds more strongly to haemoglobin, so that uric acid in the enolic form does not allow haemoglobin to transport oxygen to the periphery of the cells, to exchange it with the carbonic acid that myoglobin has taken out of the cells. If there is no oxygen, myoglobin will be left with the carbonic acid; thus, oxygen will not be available to the mitochondria. So, the mitochondria will proceed to produce the caloric energy via the fermentation of glucose, which is delivered to the mitochondria via food. The generation of caloric energy in the mitochondria by the glucose fermentation route produces lactate, but if the inner part of the cells is more acidic, lactic acid will be produced from the lactate. Lactic acid will paralyse the respiratory system of the cell, both the haemoglobin on the outside of the cell and the myoglobin on the inside.

It is a complex process, which we normally call respiration; but in reality, it is not simply an act of inhaling and exhaling air through the nose; rather, the purpose is important for all those life forms that use blood to transport oxygen into the mitochondria, and to remove carbonic acid from the mitochondria.

Chapter 3
THE EPITHELIAL LINING

If we do not correct the error of high acidity in time, the time will come when the cells become cancerous and may outnumber the healthy cells, and at this point, as magnetic beings, we will no longer be able to take control of normal respiration, because all the healthy cells will become a new kind of living cell but distorted from their original configuration.

This physical warning will happen, however, until we can observe that there is somewhere in the body a lump or tumour, which will generally be located between two layers of the epithelial membrane, because the tumour will generally occur in the apical cells. That is to say, in those cells that do not have their own blood supply, because these apical cells are supplied with nutrients from the underlying cells of the epithelial membrane that supports them. Examples of apical cells are the cells of the epidermis that are supplied with nutrients and oxygen from the lower layers of the skin, or those cells that are in the inner part (which are actually outer cells) of the milk ducts of the breasts, or the cells that line the inner part (which are actually outer cells) that form the inner layer of the seminal vesicles.

Eighty percent of cancers occur in this soft connective tissue class of epithelium, i.e., in the womb, breast, throat, liver, duodenum, colon and prostate. The lymph nodes regrow, because the lymphatic system will carry the mutated cells to the lymph nodes to destroy them. For example, the highest rate of skin cancer is in the Nordic countries where there is less sunlight, because people from those countries go down to sunbathe in the tropics. And the lowest rate of breast cancer occurs in China because Chinese women eat vegetable soy cheese instead of cow's milk cheese.

This is the main cause of cancer because we have unconsciously forced the electronic matter of the body to change to a way of functioning that is not normal for the anchorage of the magnetic mass of a human being. There-fore, it is necessary to change the way we eat in order to consciously reverse the high degree of acidity in the blood. That is to say, to make haemoglobin resume its function of carrying oxygen to the cells, but at the same time to enable myoglobin to remove from inside the cells the waste resulting from the combustion of oxygen with glucose, i.e., carbon dioxide, which was converted inside the cells into carbonic acid by the enzyme carbonic anhy-drase.

So, with a normal acidity condition, haemoglobin will take up oxygen from the lungs and bind with carbonic acid in the periphery of the cell, where it makes an ex-change with myoglobin, only when a change of acidity oc-curs from a higher value in the nucleus of the cells to a lower value of acidity of the blood in the lungs; so, that the haemoglobin releases the carbonic acid and rejoins with four oxygen atoms in the alveoli of the lungs.

It is a chemical and physically dynamic process, which, by its nature, is really the life that can occur from a chemical and physical point of view. It is really a physicochemical mechanism, which, as said, was discovered by the intuition of the German biochemist Otto Heinrich Warburg, when Dr. Warburg left a cancerous tissue in a low-oxygen, high-acidity environment to preserve the tissues and continue his research the next day. But surprisingly, Dr Warburg realised that the cancer cells had survived in a low-oxygen, high-acid environment.

The process of cellular respiration is complex; but, at the same time, respiration is important for all forms of physical life that use blood as a means of transporting oxygen and carbonic acid.

Therefore, we are obliged to know what this process of cellular respiration is like, or how it works, in order to make our respiratory system work properly, as this will allow us to live and be healthy without the delays that organic diseases bring. The lack of knowledge of how this respiratory process works is what is leading mankind into a world of wrong physical health.

If the degree of acidity is higher in the blood, i.e., if the pH is lower, as for example in the case of cancer, the oxidation state of iron will be ferric (Fe^{+3}). And with oxidation state III in iron, methaemoglobin will not transport oxygen, because haemoglobin will temporarily form a superoxide, which will not allow methaemoglobin to bind oxygen until the ferric iron III (Fe^{+3}) is converted back to its ferrous state, i.e., to iron II (Fe^{+2}).

The substance that will be responsible for reducing iron from oxidation state III back to oxidation state II and vice versa, is the ferment Nicotinamide Adenine Dinucleotide, which we abbreviate as NAD. When it is in its reduced form, NAD is written as NADH; whereas, when it is oxidised, NAD is written as NAD^+ in order to be able to identify the oxidised and reduced state of the same NAD substance.

However, it is chemically conceivable that if the superoxide associated with iron III is protonated, for example, when the blood is more acidic, the iron III in haemoglobin cannot be oxidised back to iron II. The iron in haemoglobin will remain reduced as iron III, which prevents haemoglobin from transporting oxygen from the lungs to the periphery of the cells in order to exchange it with the carbon dioxide taken out of the cell by myoglobin.

The enzyme methaemoglobin reductase will be responsible for reducing the oxidised NAD^+ back to reduced NADH, so that NAD can carry iron III back to iron II in the normal haemoglobin molecule. That is, so that haemoglobin can bind with oxygen to form oxyhaemoglobin. This is the only way that haemoglobin can transport oxygen into the cell wall via the blood vessels, so that myoglobin can bring it into the cells, making oxygen available as a fuel for the mitochondria of healthy cells.

When the iron II in myoglobin is bound to oxygen, a complex ion is formed which gives the blood a bright red colour, whereas methaemoglobin gives the blood a darker, or rather a chocolate brown colour.

The other consequence of the higher acidity in the blood is that, on the outside of the cells, this higher acidity condition will cause the antioxidant sodium urate to connect with the protons of the more acidic blood. By this effect, sodium urate will be transformed into enolic uric acid, which is exacerbated because there is also a higher carbonic acid content. In other words, with this more acidic condition in the blood, the antioxidant sodium urate will be converted to enolic uric acid, but not to ketonic uric acid. Ketonic or enolic uric acid will not function as an antioxidant, because it does not have the charges to absorb the free radicals that result from haemolysis; that is, ketonic or enolic uric acid is a neutral compound that does not function as an antioxidant; because, precisely, sodium urate is converted to uric acid that should be ketonic uric acid, after sodium urate has fulfilled its function as an antioxidant in normal blood.

Uric acid that is in excess in the more acidic blood will be in the enolic form. Normally, uric acid should be in ketone form. Uric acid in the ketonic form will not attack cartilage calcium or bone calcium, but uric acid in the enolic form is a stronger acid than uric acid in the ketonic form; therefore, uric acid in the enolic form will attack joint cartilage calcium leading to arthritis, and uric acid in the enolic form will attack bone marrow calcium leading to osteoporosis.

Under normal blood acidity, uric acid should remain in its ketonic form; that is, at a normal blood pH of 7.40, the uric acid form is ketonic. But if the degree of acidity of the blood is higher, i.e., below a pH of 7.40, the uric acid will be in its enolic form. The enolic uric acid will attack the calcium in the bone marrow, but it is precisely in the bone marrow that red blood cells are formed, resulting in decreased haemoglobin levels in the blood and osteoporosis. Additionally, uric acid in an enolic form will cause weakening of the collagen fibre, resulting in deformed or rheumatoid arthritis. The lack of collagen will cause the skin in diabetic wounds not to bind together, as in the case of people with so-called diabetic foot.

The calcium released from the cartilage and bone marrow will form calcium urate crystals with the enolic uric acid. These crystals act with a sharp effect between the bone and the muscles and cause the intense muscle pain known as fibromyalgia, i.e., pain in the muscles attached to the bones.

This explains why terminally ill cancer patients have up to a thousand times more acidity in their blood than normal. Thus, healthy cells are forced to mutate, i.e., to survive without oxygen, as long as we, as spirits, do not decide to make a change in diet to reverse the body's high acidity. For low acidity with alkalinity in the normal range is a chemical condition necessary for the survival of healthy cells.

If such a change of acidity is not made in a timely manner with the diet, the surviving cancer cells will grow in number; thus, the healthy cells will come into conflict with the cancer cells. Healthy cells need oxygen and low acidity to live, but cancer cells do not need oxygen and can survive in a higher acidity environment. As we saw, low acidity is a basic chemical condition for oxygen transport. These mutated cells can survive, but by breathing in a different way than healthy cells.

Cancer originates chemically, from the moment when, in the nucleus of healthy cells affected by high acidity, the bases guanine and uracil undergo tautomerism and methylation. As spirits made by magnetic mass, we will have to leave that body made by electronic matter, until we have the opportunity to board another physical body that is healthy.

However, as we envision it, the process of gestation and birth is also a complex event, as it depends on several factors: mainly the factor of finding conscious parents to guide us in order to help our birth to occur and then how to live without the diseases.

Some children are born vegan, but the carnivorous parents think that the child is abnormal because the child does not eat meat; because, for the carnivorous parents, the normal thing is that the child eats meat; so, the carnivorous parents will try to convert what is normal for them; that is, that the vegetarian child becomes a predator who kills his genetic brothers and sisters to devour them.

So, all human beings, without exception, should be aware of how the process of respiration and the chemical functioning of their cells works.

Those who are unaware of the terminal stage of cancer should also avoid consuming dairy products, because the fats in these foods will increase the viscosity of the blood. Thus, the blood will flow with greater difficulty to the cells, resulting in increased blood pressure, and internal strokes may occur in the most fragile blood vessels, which are precisely the blood vessels in the affected epithelial tissue. Such as in the tissue of the prostate, breast, duodenum, and colon.

One of the most influential factors for cancer to occur in the epithelial tissue of the prostate is the Escherichiae Coli bacteria, which comes along with unhygienically handled meats. Because there was a migration of E. coli bacteria from the viscera into the meat of the slaughtered animal. E. coli bacteria in soft tissue, whether in the pancreas, womb, breast, or prostate, generate carbon dioxide, and the carbon dioxide will be converted into carbonic acid by the enzyme carbonic anhydrase, which will increase the degree of acidity within these areas of soft tissue, i.e., tissue or epithelium membrane. But no blood will reach these parts of the body to regulate the high acidity, as explained in the case of the apical cells. So, the carbonic acid that is not being removed will corrode the collagen fibre of the prostate and milk ducts, thereby increasing the rate of cancer in these epithelial tissues. It would be wise not to kill the animal.

In fact, breast cancer begins within the milk ducts, whereas prostate cancer must begin in the seminal vesicles, and from there, the cancer will begin to spread to the prostate, because E. coli bacteria can lodge in the seminal vesicles of the male reproductive system, and because they are hollow structures, as mentioned, blood with antibodies will not reach these vesicles.

The consumption of animal protein is also affected because animal protein contains methionine. As mentioned, methionine is the most abundant amino acid, because methionine is contained in all animal proteins.

The only option left to the cells that remain healthy but immersed in this cancerous situation, or to survive alongside these cancerous cells, in order for the healthy cells to be able to capture a greater quantity of haemoglobin with oxygen, i.e., oxyhaemoglobin, is to create a greater number of blood vessels around them; which will aggravate the amorphous appearance of this mixture of the two kinds of cells, due to a greater amount of unconnected epithelial mass that will form between two sheets of epithelial tissue, but now this deformation will appear differently in the form of a polyp or tumour.

The pH value of the urine and saliva of those patients in the terminal stage of cancer is in the order of 5.5 and 4.5; that is, the blood of these people is more acidic, taking as a reference the blood of a healthy person. Therefore, the normal process of cellular respiration and anti-oxidation will be more difficult for these people, but in addition, people involved with cancer will not be able to achieve haemolysis of the red blood cells that have ceased to function as transporters of oxygen and carbonic acid.

We are going to see mathematically, what is the range or value of the concentrations of sodium urate and ketonic uric acid, to demonstrate why and how it is that sodium urate is converted into ketonic uric acid, which is a consequence of the mutations that occur in the cells. By means of this relationship, we can see in what proportions ketonic uric acid and sodium urate are found in the blood of a normal person and in a person with cancer. This can be deduced from the following mathematical relationship of approximate solubility, which shows the ratio of the concentration of sodium urate to the concentrations of ketonic uric acid and enolic uric acid:

$$[\text{sodium urate}] = 10^{(pH-pka)} \, [\text{ketonic uric acid}]$$

The pk_a of ketonic uric acid is 5.8; therefore, substituting the values for the pH value of a person whose blood has a normal acidity value or pH equal to 7.40 we have that:

$$[\text{sodium urate}] = 10^{7.4-5.8} \, [\text{ketonic uric acid}]$$

$$[\text{sodium urate}] = 1 \, 0^{1,6} \, [\text{ketonic uric acid}]$$

$$[\text{sodium urate}] = 40 \, [\text{ketonic uric acid}]$$

In other words, for the blood of a person whose blood acidity value is normal, the sodium urate concentration should be approximately 40 times higher than the ketonic uric acid concentration.

Whereas, for the blood of a person involved with a terminal case of cancer, the pH of the blood is 5.5; so, in this relationship if we assume that the pk_a of the enolic uric acid is equal to the pk_a of the ketonic uric acid, we have that:

$$[\text{sodium urate}] = 10^{5.5-5.8}\,[\text{enolic uric acid}]$$

$$[\text{sodium urate}] = 10^{-0.3}\,[\text{enolic uric acid}]$$

$$[\text{sodium urate}] = 0.5\,[\text{enolic uric acid}]$$

Which indicates that, if the blood is too acidic for a person with a terminal case of cancer, the concentration of enolic uric acid doubles, i.e., the concentration of enolic uric acid is twice as high as the concentration of sodium urate:

$$[\text{enolic uric acid}] = 2\,[\text{sodium urate}]$$

That is, in the blood of a person with end-stage cancer, there will no longer be the antioxidant sodium urate, or perhaps any other antioxidant available, to reduce the iron in haemoglobin from ferric ion III to ferrous ion II. Most likely, this high acidity in the blood will also affect the antioxidants NAD^+ and NADH. Because, if the acidity value in a person with a terminal case of cancer were 4.5, the ratio [sodium urate]/[enolic uric acid] would be reversed. A person with a terminal case of cancer, the uric acid will be in the enolic form; therefore, it becomes necessary to reverse this process of higher acidity in the blood.

At this stage of terminal cancer, the concentration of enolic uric acid would be 2 times higher than the concentration of sodium urate. So, all healthy cells will be starved of oxygen; thus, the cellular body of the person with terminal cancer stops functioning due to lack of oxygenation. But in that final stage, the lack of red colouring of the blood will cause the person's countenance to be anaemic or yellowish in colour because of the increased concentration of methaemoglobin.

Thus, the problem of acidosis reached its climax; because even when cancer invades the body in the form of metastasis, the acidity reading may be among those higher acidity values. An acme or paroxysm will occur in the person with the most acidic blood environment, where the rest of the healthy cells collapse, as they will not be able to breathe to take up oxygen and survive. Whereas the cancer cells changed the form of existence of the healthy person, forced by the spirit of the magnetic mass inhabiting a body made of electronic matter, which was not configured to eat meat but vegetables. Vegetarians do not have in their catabolic system the enzyme urate oxidase, which would allow them to convert sodium urate to the form of allantoin in their kidneys and urinary bladder instead of uric acid.

Cells can be modified because the electronic matter in the cells that make up the physical body is condensed electronic energy; that is, electronic matter is the only kind of integrated electronic energy that is prone to undergo physical changes in space over time.

Because of the high acidity, organic conditions have been reached, so that both kinds of healthy and mutant cells cannot coexist occupying the same electronic body. These conditions of higher acidity in the blood are favourable only for the survival of the mutant cells because these changed cells can survive without oxygen. Because if there is no oxygen, this situation is not favourable for those cells that are still healthy. This will happen until the anomaly of high acidity, which was caused by the imbalance as a consequence of the lowering of the pH value, i.e., the high acidity of the body, is reversed in time. Until we find a way to lower the acidosis, we have no other way to reverse the cancer.

A successful strategy of some people with cancer is that they have changed their lifestyle from carnivores to vegetarians, and have been relieved of the disease, even in those with end-stage cancer. Perhaps it is that, with this timely change of dietary strategy, they managed to restore the blood that had become acidic or to return to its normal acid value. Perhaps it is because these cancer patients have understood in time that it is the consumption of meat that causes the damage, which contains the cells that cause acidity and then tautomerism. Meanwhile, the proteins in meat also induce methylation of the cytosine and uracil bases, which causes the cytosine and uracil bases to convert to the thymine base.

The only way to give the normal cells a new chance is for them, on their own autonomy, to regain control of their chemical balance, or the proper condition of their functioning, perhaps trying as far as possible not to force them back into the conflicting problem of becoming cancerous again.

Knowledge will be what will prevent us from going down this wrong path of nourishment again. Hopefully, this knowledge will reach all human beings, so that a more conscious race will definitely be generate or that it will conform to a new kind of humanity, for the magnetic mass of the spirit and the Universe will always remain eternal. So, we still have a long way to go and must learn how to evolve in this vast Universe, but it will be an equal condition for all beings existing in the Universe, for the right to live on Earth is not a form of life that is exclusive to human beings.

We conclude that the origin of cancer is due to an acid-alkali imbalance in the blood, which can be reversed chemically but not with a vaccine. Because the case of cancer is not an immunological but a chemical problem. The pathological differences in this abnormality are due to the kind of epithelial tissue involved. There will additionally be a problem known as anorexic cancer, which manifests itself in those with end-stage cancer as a reluctance to move forward on the right path. Since, in that advanced stage of cancer, there will be an increased lack of appetite, and the hopelessness due to the lack of oxygenation will make the affected person feel without energy.

So, the cancer sufferer will fall into a state of sleep more often than normal in order to take up more oxygen, and the lack of oxygen will become the main cause of the body's lack of spirit, but perhaps not because of the cancer itself, but because the lack of interest in food and the hopelessness of suffering from the cancer disease will create this indisposition or listlessness, which will worsen the health aspect of the cancer sufferer.

Additionally, the second most important antioxidant in the blood after sodium urate is vitamin C, and because it is water-soluble, we lose vitamin C every day through urine and perspiration. Therefore, we will have to obtain vitamin C from fruit consumption as a source of antioxidants.

We will not need to consume the cells of another animal to obtain sodium urate from them, as we will obtain this antioxidant from the cells that die; for it is from the adenine and guanine purine bases of our DNA and the various extinct RNAs that we will obtain the necessary amount of our main antioxidant sodium urate.

It is by a seemingly insignificant change in the acidity value between the renal fluid and the blood that a proper balance of both sodium urate and ketonic uric acid is maintained; which has to move within a narrow concentration range; and this is determined by a constant called the dissociation or chemical equilibrium constant, i.e.:

Keq= [sodium urate] x [protons]/ [ketonic uric acid]

The concentration of sodium urate in the blood is:

[sodium urate] = K_{eq} [ketonic uric acid] / [protons]

The amount inside the brackets is read as the concentration.

As we have seen, in the blood of a healthy person, the sodium urate concentration has to be 40 times the value of the ketone uric acid concentration. This means that the dissociation constant of the ketonic uric acid Keq in the blood must be very large, i.e., the ketonic uric acid must be almost completely dissociated in the form of sodium urate for the proton concentration to remain constant. But the dissociation must be within a narrow range of pH values, because this range must not be above 7.45 and not below 7.35, i.e., in reality this pH value of normal blood must be around 7.40. If this pH value is 7.35, the problems of acidosis arise. Whereas, if the value is at a pH value of 7.45, the problem of alkalosis will arise.

Both problems acidosis and alkalosis, are determined by the value of this equilibrium constant, which is related to the concentration of protons in the blood. If the value of the proton concentration moves towards higher values, the chemical equilibrium will also move, i.e., K_{eq} is constant in order to keep the concentrations of sodium urate and uric acid in the new range. However, the concentration of uric acid will become larger; therefore, the ketonic uric acid will be in its enolic configuration. These pH values are only determined by the acidity range of the blood.

So, the cancer problem can of course be reversed chemically as soon as we can lower the concentration of protons in the blood. If somehow, we were able to keep this balance within the range at which cells function normally, then of course cancer would not occur, because there is no chemical reason for cancer to occur in humans.

Chapter 4
TAUTOMERISM

Tautomerism refers to a change in the electronic con-figuration of a ketone to become an alcohol, as can be seen in Figures 3, 5 and 8. When tautomerism happens to a ketone, it will cause the couplings between the base pairs to change; therefore, the electronic structure of DNA will also change. In DNA, the two ketone bases that can undergo tautomerism are the guanine base G and the uracil base U in Figure 7, because the guanine base has an alpha hydrogen on nitrogen 1, while the uracil base has an alpha hydrogen on nitrogen 3. The hydrogen that is in the alpha position or closer to the carbonyl group of the ketone (>C=O) is more probable to come out. When the medium is acidic, the alpha hydrogen is released and the carbon atom of the carbonyl becomes positive; thus, the double bond on the positive carbon of the carbonyl group is closed. The carbon atom can only connect to 4 atoms, so the oxygen in the ketone becomes negative and bonds with the alpha hydrogen, which is released. Thus, the ketone becomes an alcohol. The guanine base can change from its normal ketonic form to its alcoholic form. This is basically what happens electronically with tautomerism of a ketone that has an alpha hydrogen nearby, i.e., a hydrogen atom that can leave.

Tautomerism cannot happen to the thymine base T, which has an alpha hydrogen on nitrogen 3, because the thymine base has a methyl group on carbon 5, so the thymine base is aromatically stabilised. That is, the double bond is displaced around the thymine base ring.

The uracil base undergoes tautomerism after it has lost its beta hydrogen, i.e., the uracil base can undergo methylation after it has undergone tautomerism. Thus, the enolic uracil base will also become the thymine base. As shown in Figure 13.

Tautomerism of both the guanine base and the uracil base can rearrange the shapes of their couplings under the effect of acidosis, which is essentially an electronic process. In the methylation process, both the uracil base in enolic form and the cytosine base will be converted to the thymine base. So, to form DNA, the chromosomes will continue to pair the adenine base with the thymine base; but, once the cytosine and uracil bases are removed from the nucleus, the chromosomes will have to pair the guanine base in its enolic form with the thymine base. This new DNA will be in error with respect to its original electronic configuration; that is to say, it is a DNA that does not correspond to the original DNA, which configured the human being before its bases underwent the process of tautomerism and methylation. As a consequence of the increase in the degree of acidity in the nucleus of the cell, a cellular mutation of the DNA will occur, which will manifest itself as a tumour.

The cause of tautomerism comes from the consumption of cells of animal origin because the adenine and guanine purine bases of the DNA of the ingested cells will be converted into sodium urate in parallel. For the meat cells of any living being are chemically the same as the cells of a human being. So, when these dead cells that come with the meat of any animal are ingested, the adenine and guanine bases of the meat's DNA will also be converted into sodium urate.

However, in the blood of a human being, in order to maintain a chemical equilibrium between the concentration of sodium urate and the concentration of uric acid in ketonic form, the excess sodium urate i.e., the sodium urate coming from the consumed cells, will also have to be converted into ketonic uric acid; but, that excess ketonic uric acid will be converted into enolic uric acid, according to the following chemical equilibrium equation:

[ketonic uric acid] + [enolic uric acid] ↔ [sodium urate] + [protons]

This means that, when there is a high concentration of sodium urate in the blood, in order to maintain the balance in the above ratio, the concentration of ketonic uric acid must increase. But a part of the ketonic uric acid will be converted into enolic uric acid, in order to maintain a balance between all the electronic species, i.e., between the concentration of ketonic and enolic uric acid and sodium urate and there will be more H^+ protons in the blood and the high acidity of the body will be produced.

Whereas the high concentration of H^+ protons on the right will reach a point where it can no longer be regulated in normal blood. That is, by the sodium bicarbonate-carbonic acid regulating system, whose buffering capacity controls the blood acidity value to remain within its normal functional range, which should be maintained at around a pH value of 7.40.

If an increase in acidity occurs, the balance will be forced to move into a higher range of H^+ proton concentration, and ketonic uric acid will become enolic uric acid as the regulatory capacity of the sodium bicarbonate-carbonic acid system is lost.

The body's acidity regulating system is also known as the buffer system. The sodium bicarbonate to form the buffer came from sodium chloride consumed with food, when the sodium chloride salt reacted with hydrochloric acid in the stomach, and sodium bicarbonate was formed by the action of the hormone secretin. Carbonic acid, on the other hand, was formed from carbon dioxide by the enzyme carbonic anhydrase.

In bile, sodium bicarbonate is used to neutralise the stomach acid carried by the chyme when the chyme leaves the stomach at the level of the duodenum. Chyme is acidic because it is stomach acid that activates pepsin to break down ingested proteins. When there is no hunger, the enzyme pepsin does not attack the stomach-forming proteins, because pepsin is in the form of pepsinogen, which is activated as pepsin when the acidity of the stomach is increased by stomach acid. However, when hunger is not satisfied in time, pepsin can attack stomach proteins, causing stomach ulcers that begin in the epithelial tissue of the duodenum. In fact, stomach cancer originates in the epithelial membrane of the duodenum.

In other words, the degree of acidity in the small intestine from the duodenum onwards must be alkaline so that carbon dioxide gas bubbles do not form from the sodium bicarbonate, which causes refluxes that can lead to belching and the dragging of bile fluids into the oesophagus, i.e., gastritis. Similarly, in the duodenum, the acid chyme coming out of the stomach must be completely neutralised, because the small intestine contains enzymes called trypsin, chymotrypsin, and the enzyme pancreatic lipase. The enzymes trypsin and chymotrypsin are activated at an alkaline pH; therefore, the chyme is neutralised in the duodenum. Because they are deactivated as trypsinogen and chymotrypsinogen, the enzymes trypsin and chymotrypsin do not attack proteins in the small intestine. When active, trypsin and chymotrypsin break down protein residues carrying aromatic amino acids. Meanwhile, the enzyme pancreatic lipase converts fats and oils ingested with food into fatty acids.

Therefore, as more meat is consumed, more cells and more saturated fats will be consumed; and with more cells consumed, the amount of adenine and guanine purines will increase in the body as the age of the human individual increases. The amount of meat consumed will be proportional to the body weight of the human being who wants to be and eat like a carnivore. In such a way that the acidity of the blood of a human being who wants to be a carnivore, the acidity of his blood will gradually fall out of its functional range, and thus the blood pH will decrease, i.e., the acidity of the blood of the human being who imitates carnivores will increase.

Humans are vegetarians because they do not have the enzyme urate oxidase in their catabolic system. The enzyme urate oxidase converts sodium urate into allantoin. Carnivorous animals by nature have the enzyme urate oxidase in their catabolic system; therefore, carnivorous animals have more acidic blood than humans. The enzyme urate oxidase allows carnivorous animals to convert the cellular excess of consumed meat cells to allantoin instead of uric acid.

When cells and saturated fat from the meat of a sibling animal are consumed, glial cells that are linked to neurons can be affected and neurological diseases, such as weakness, anxiety, Parkinson's, or Alzheimer's, will be induced.

But it doesn't matter what kind of animal meat you eat, because all animals are made up of cells: whether you call them cow, sheep, chicken, or fish, they are all living beings that are made up of cells. Apart from what we have said, all these living beings are our genetic and energetic brothers and sisters.

However, one of the biggest problems for a human being who wants to become a carnivore is that the enolic or ketonic uric acid is insoluble when the blood fluid is more acidic. Because the only fluid that has to be acidic in the body is the fluid in the kidneys and urinary bladder. Therefore, uric acid with excess; that is, uric acid that was formed from sodium urate from the adenine and guanine bases of meat cells will be in the enolic form, but not in the ketonic form. As stated, the ketonic form of uric acid is weaker than uric acid in the enolic form; therefore, uric acid in the enolic form will deprive haemoglobin from transporting oxygen from the lungs to the periphery of the cells; thus, the cell will be forced to produce heat by glycolysis of glucose; which generates lactic acid inside the cell. Lactic acid is stronger than enolic uric acid and will produce tautomerism of the guanine base in the DNA of the cell nucleus and cancer ensues. Lactic acid is used in cosmetics as a skin exfoliant. When enolic uric acid accumulates in the blood, it begins to release calcium from the bones, and calcium urate will form; but, once the calcium urate passes through the acidic environment of the kidneys, the calcium urate will crystallise, and kidney stones will form. The same is true for the formation of the enolic uric acid crystals that form gallstones in the gallbladder. It is in the enolic form that crystals are extracted from the joints of arthritic people. The two crystals are easy to identify; ketonic uric acid crystallises in the form of discs, while enolic uric acid crystallises in the form of needles.

In an imperceptible way, acidosis is created, which, on the other hand, will create the problem of tautomerism in the guanine and uracil bases. Because the protein that is consumed together with the piece of meat certainly provides us with proteins, but these meat proteins bring with them an excess of the amino acid methionine, which, on losing its methyl group, will cause the methylation of cytosine and the methylation of the enolic uracil.

As mentioned, the uracil from its ketone form will move to its enolic form; and from the enolic form the uracil as well as the cytosine base will undergo a methylation process and the result of this methylation process is that both cytosine and uracil bases will be converted to the thymine base. One situation is a consequence of the other, which is why we have used the extensive symbol & in the subtitle as: 'Tautomerism & Methylation', instead of 'Tautomerism and Methylation'. So, diseases exist because we cause changes in our DNA through the way we eat.

The guanine and adenine bases of the DNA of cells that cease to function in our body system are converted to sodium urate. However, when we ingest animal meat cells, the ingested adenine and guanine bases will also be converted to sodium urate, but in order to maintain chemical balance, the sodium urate consumed will be converted to ketonic uric acid and ketonic uric acid in excess will be converted to enolic uric acid. The enolic uric acid will cause damage to our cellular system, as such excess cell consumption will create acidosis. Just because we want to be carnivores, we are consuming a class of cells that are chemically the same as ours; that is, these ingested cells contain the same DNA and RNA as ours. Because the only thing that makes us different is the position of these stretches of bases in each DNA and RNA, which gives us a different configuration through the genetic code; but the genetic code is unique to each species and to each individual of each species.

We can say that the tautomerism that occurs to our guanine and uracil bases is a consequence of acidosis, since acidosis induces a mutant error of the chromosomes in the DNA, because tautomerism alters the forms of the coupling of the bases in the DNA and RNA. Although this fact is verifiable, since, as mentioned, it is in the enolic form that uric acid crystals are obtained in the joints of people with arthritis. To be more specific, this uric acid in the joints of arthritics and that found in the kidneys is actually in the form of 3-methyluric acid; that is, it is uric acid in the enolic form.

The process of tautomerism is chemical, so we have no other way to explain it. But, as we have said, the phenomenon of tautomerism occurs when the carbonyl group of a ketone becomes an alcohol; since, in an acidic medium, alcohols are more stable than ketones.

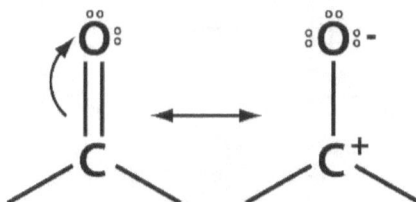

FIGURE 3
HIGH ACIDITY, CAN CONVERT THE CARBONYL GROUP OF A KETONE TO AN ALCOHOL

So, although the double bond of the ketone (>C=O) on the left of Figure 3 is stable, ($\approx$178 kcal/mol) it is only slightly stronger than the single bond (>C+ –O⁻) of the alcohol shown on the right of Figure 3; therefore, the amount of energy is approximately 171 kilocalories per mole (2 x 85.5 kcal/mol).

We can say that the energy contribution of the form on the right in Figure 4 below can in some cases be as much as 50% of that on the left, which means that it is possible that both ketone and enolic electronic forms can coexist independently to form two distinct compounds, i.e., a ketone in equilibrium with its alcohol.

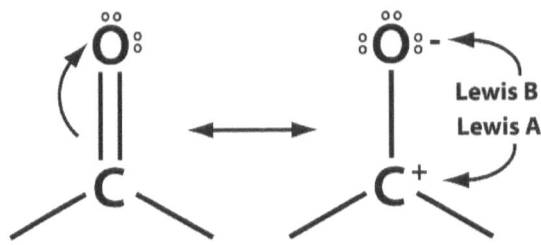

FIGURE 4
BEHAVIOUR OF THE CARBONYL GROUP AS AN ACID AND AS A LEWIS BASE AT THE SAME TIME

When defining the concept of pH, an important clas-sification of ionic reactions in organic molecules is based on the nature of the reactive particle, which is conven-iently assumed to be the attacking species. From this point of view, according to the Gilbert Newton Lewis definition, an acid is a species capable of accepting a pair of electrons since its electronic charge is positive. While a Lewis base is a substance that gives up a pair of electrons because its electronic charge is negative.

With this definition thus established, we have that: those organic substances that are electron acceptors are called Lewis's acids; therefore, they are identified as electrophilic substances; that is, electrophilic species are those substances that have an affinity for particles that have a negative charge. Meanwhile, the electron donors are the Lewis bases, and they are called nucleophiles, since they are particles that have an affinity for electronic nuclei.

This is how organic reactions are generated, which are classified as electrophilic and/or nucleophilic, depending on the type of electron-giving or electron-accepting reagent that gives rise to these chemical reactions. Therefore, we deduce that the carbonyl group of a ketone from which the conditions can be given for equilibrium to be formed with its alcohol will behave simultaneously in the same molecule as an acid, but at the same time as a Lewis base or alkali, such as the forms shown in Figure 4, thus establishing a keto-enolic equilibrium.

This is an inherent property or characteristic of the behaviour of the carbonyl group of a ketone that has an alpha hydrogen, since the electrons tremolate from one form to the other in those compounds where this possibility exists or depending on the degree of acidity. In other words, the compound will behave like an acid and a Lewis base at the same time. This means that these substances will behave as acids or alkalis depending on the acidic conditions of the medium in which they are immersed. Substances that have these characteristics of behaving according to the degree of acidity are called amphoteric.

Thus, if the substance behaves like a base, it will capture a Lewis acid, as happens with the bases guanine and uracil, which can capture a proton (H^+) in their carbonyl group from the acidic medium, if they are behaving like an alcohol instead of being ketones when the environment of the cell nucleus becomes more acidic. In this case, it is the internal fluid of the cells affected by acidosis, which will influence the acidic conditions of the antioxidant system within the cells. Mainly NADH and NAD^+, which, as we have seen, is responsible for oxidising iron II in haemoglobin to iron III, and reducing iron III back to iron II, so that haemoglobin can transport oxygen as iron II and carbonic acid as iron III. To keep the body system within its functional acidity range, an antioxidant system is needed within the cells.

If these high acidic conditions occur within the cells, the carbonyl group of the ketone $>C=O$ will be transformed into an alcoholic group, $\equiv C\text{-}OH$. So, if the intercellular medium becomes acidic, the ketone, or Lewis's base, will be transformed into an alcohol; that is, it will be transformed into a Lewis's acid, which is more stable and reactive than the ketone when the medium becomes more acidic.

Under these circumstances, molecules in this situation would be forced to undergo internal regrouping in the chromosomes in the nucleus of the cell, as happens to a ketone, or which was forced to transform into an alcohol. More stable enolate ions can form, such as those shown on the right of Figure 5.

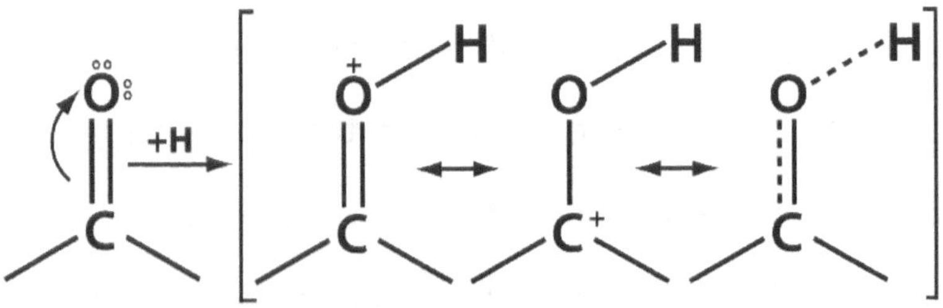

FIGURE 5
FORMATION OF AN ENOLATE ION FROM LEWIS ACID ON THE CARBONYL GROUP OF A KETONE

In the opposite process, on the enolate ion on the right of Figure 5, if protonation were to take place on carbon, in this case, the ketone would regenerate again. But, if protonation occurs on oxygen, an enol ($\equiv$C-OH; -keto, -OH, -enol) will be formed. So, a ketone with these changing characteristics, or having an alpha hydrogen, will always be in equilibrium with its enol, which will depend on the acidic conditions in the cell nucleus, as shown in Figure 6. C is the ketone; HA is the alpha hydrogen and E is the enol.

However, as we can see, there can be intermediate states, which can be seen in Figure 5. For the case of tautomerism on the enolic base uracil, if there is a nearby group that partly satisfies the positive charge that is generated on the carbon, the acid will be relatively less acidic, i.e., more basic, because the acidity moves on a relative scale between 0 and 14. But it is assumed that, at pH 7.00, the degree of acidity is neutral, although that or any point in equilibrium is difficult to achieve, since pH 7.00 is really a transitional state between acidity and alkalinity.

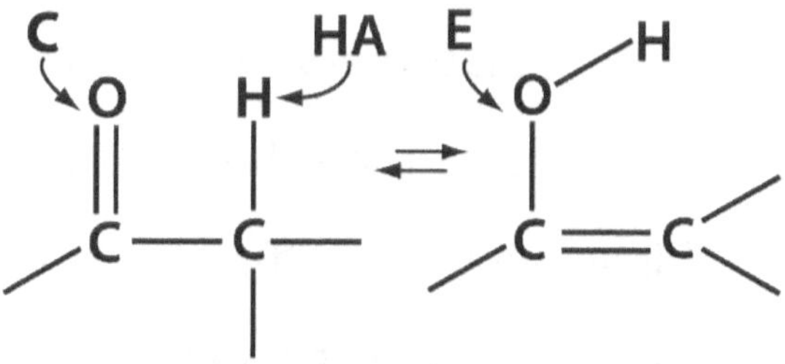

FIGURE 6
KETO-ENOLIC EQUILIBRIUM BETWEEN A KETONE AND ITS ALCOHOL

An important feature is that the ketonic and enolic forms are real molecules, i.e., they are independent and distinct molecules and should therefore not be confused with resonance isomers, which are theoretical intermediate forms that are highly reactive, but do not stop to form stable substances with a real physical existence. For example, it is possible to prepare enolates in the laboratory.

In order to identify or describe the relationship between the ketone and enolic forms, another name has had to be adopted: they are called tautomers and when these electronic inter-conversions occur between keto-enol forms or from a ketone form to an enolic form, the phenomenon is known as tautomerism. Tautomer is derived from the English word taut.

At equilibrium, tautomers are formed, but quickly switch from one form to the other, even under ordinary conditions. This is why it is difficult to isolate them for characterisation or identification in the laboratory; it is like telling the pH at the 7.00 point. But for the same reason, it would be impossible from a practical point of view to measure this keto-enolic balance in the blood of a person suffering from cancer. At least to be able to prove that this is the cause of the cancer, or to prove the existence of these ketone and enolic compounds as two distinct and independent substances. If one wants to explain the phenomenon of tautomerism, it becomes obvious and reasonable from the point of view that it is deduced by electronic and theoretical analysis of the molecular structure of each molecule that is prone to participate in a tautomerism process.

As it is impossible to measure, for example, the degree of displacement of the tautomeric equilibrium of a DNA in vivo; this equilibrium has been attempted to be measured by performing in vitro experiments using the so-called 'Combined Density Functional Theory' with the Poisson-Boltzmann continuous solution model. But this is a theoretical method, which will lead to a theoretical probability by experimental simulation. But it is what we can deduce in theory, just by sharpening the analysis, and knowing the chemical characteristics of the five bases that make up the DNA and RNA of the cells shown in Figure 7.

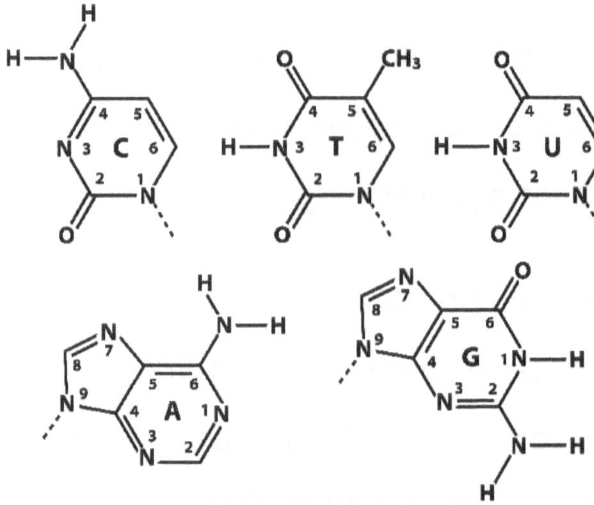

FIGURE 7
THE FIVE BASES THAT MAKE UP DNA AND THE VARIOUS ARNS IN THE NUCLEUS OF CELLS

Chapter 5
DNA BASE LINKAGE

Having looked at how tautomerism can happen, let us now review to see how base pairs are wrongly connected in DNA. Then, we can see this analogy with the chemical problem that we have induced on our own due to a bad feeding strategy; because, by the way we feed ourselves, we induce a change in the shape of the base pair couplings that make up the chromosomes that synthesise normal DNA, as shown in Figure 10. It is basically by tautomerism and methylation that causes the mutation that will manifest itself in the physical body as cancer.

In Figure 7, we can see the five bases that are in the nucleus of the cells for the chromosomes to build the normal DNA sequence. Four of these five bases are making up the DNA. The four bases involved in forming normal DNA are: adenine, ketonic guanine, thymine, and cytosine. The linking groups that are not shown are the dashed lines (---) that correspond to the deoxyribose sugar molecules that form the side chains of DNA and is what we identify as a nucleoside. In messenger RNA, there is uracil, the fifth base of this group.

To build these bases in the nucleus, cells need folate as a raw material to form folinic acid. Folate is found in green fruits; it is what gives fruits their growth. One of the active forms of folate is folic acid, which is why folic acid intake is recommended during pregnancy to prevent genetic errors in the embryo, such as bifid or open spines.

We can say that, under the strictest conditions of acidity or the normal chemical environment inside the nucleus of the cells, in the chromosomes, the thymine base is achieved by participating only in the DNA, but the thymine base does not participate in the conformation of the messenger RNA. Whereas the same is true for the uracil base, which is present exclusively in the various messenger RNAs, but uracil is not in normal DNA.

Given that in DNA there is only the thymine base but no uracil, or that the uracil base is only in messenger RNAs but not in normal DNA, this means that under the natural conditions of cell functionality within the nucleus, chromosomes can only use the nucleophilic or negative bases, in order to connect the base pairs via hydrogen bridges, with the electrophile or positive acid groups, in those sections where the rows of normal DNA and the different messenger RNAs are closed. That is, the messenger RNAs that are derived from normal DNA.

It means that, somehow in the nucleus of cells, the bases that make up the normal DNA are prone to changes, which happen according to the acidic or basic conditions of the cell nucleus. It is the basicity, or acidity, that determines the shape of these normal hydrogen couplings between base pairs in chromosomes. Therefore, the acid-base conditions for the couplings to occur will be determined by the degree of acidity that prevails at any given moment within the cell nucleus; since, as can be seen, this very specific way of mating the bases by means of hydrogen bonds depends on the functions that each pair of bases must fulfil in the normal DNA and in the messenger RNA, inside and outside the cell nucleus; but always within the capsule that makes up the structure of the cell.

If we look at Figure 7, we notice that the only thing that differentiates the thymine base from the uracil base is that the thymine base has the methyl group (-CH$_3$) inserted on carbon 5 of the ring. In such a way that, somehow, or because the methyl group is a reactive negative charge-giving species, the methyl group, being close to the ketone group of the thymine base, stabilises the thymine base ring in an aromatic form; therefore, this methyl group does not allow tautomerism to occur to the thymine base. Nor can methylation occur to the thymine base. In other words, the methyl group does not allow the thymine base to become an enol.

The other reason is that, at carbon 5, the methyl group replaced the alpha hydrogen, so thymine cannot undergo tautomerism. The thymine base has a beta hydrogen at carbon 6, but the thymine base is less likely to be tautomerised. Whereas, relatively speaking, tautomerism is more likely to occur with greater intensity in the ketonic guanine base, because the oxygen in the ketonic guanine base will attract the proton from the acidic medium or nitrogen that is adjacent to the ketone group, i.e., nitrogen number 1.

After tautomerism occurs to the uracil base, we can observe that this base has two alpha hydrogens adjacent to the carbonyl group on carbon number 4, specifically on nitrogen number 3 and carbon number 5. In such a way that a double bond can form in the uracil when the enol is produced by the exit of the hydrogen from carbon number 5, since this double bond can be delocalised by rotating in the aromatic ring; but then the alpha hydrogen can leave more easily, which is on nitrogen number 3; which is more likely to happen on the uracil base, when the liquid medium of the cell nucleus is more acidic.

The adenine and guanine bases correspond to the purine group, i.e., they are less basic bases. Whereas the bases cytosine, thymine and uracil belong to the pyrimidine group. It means that these bases are more basic. When we say that these bases are more basic, we can point out that, as we have said, although all four substances are bases, the Lewis definition of base and acid is also a relative question, since a base or an acid does not define a definitive character or condition. We can identify that one acid is less or more acidic than another, or that one base may be in relative terms more basic than another, and so on. But in any case, it would be difficult to define or know where the concept of base ends and where the definition of acid begins or vice versa, since it would be difficult to locate the neutral point of acidity, or the neutral point at which the pH is exactly 7.00.

According to what we have seen in this keto-enolic equilibrium, those bases that contain in their electronic structure ketone groups ($>C=O$), plus an alpha hydrogen, which can be detached, these bases can be configured in

the form of an enol; that is, an alcohol ($\equiv$C-OH) so that double bond is produced in the ring; in such a way that when this base becomes an enol, the molecule is more aromatically stable.

Whereas, the cytosine base, despite having a ketone group on carbon 2, this pyrimidinic base has the main characteristic of not having an alpha hydrogen on the adjacent nitrogen on carbon number 1 and 3 of its ketone group. In other words, the cytosine base does not have an alpha hydrogen that can be detached to capture one of the bonds and then stably close the double bond that gives it its aromaticity, which is a necessary condition for the enol to be formed.

The double bond in the ring of the cytosine base is complete with hydrogen atoms, so that the cytosine base is unalterable so that an electronic tautomerism process can occur.

We conclude that what can happen to the cytosine base is demethylation when the cellular environment becomes more acidic, as the high acidity will cause carbon 5 of the cytosine base ring to become exposed to nucleophiles or nucleosome scavenging groups, such as the methyl radical, when the cellular environment becomes more acidic, or because these methyl groups, as we have seen, are in abundance due to the consumption of animal protein. This leads to the demethylation of the amino acid methionine, but the amino acid methionine is the amino acid that is found in the greatest abundance in all proteins of animal origin, since methionine is the amino acid that marks the starting point for the synthesis of the protein chain by the ribosomes.

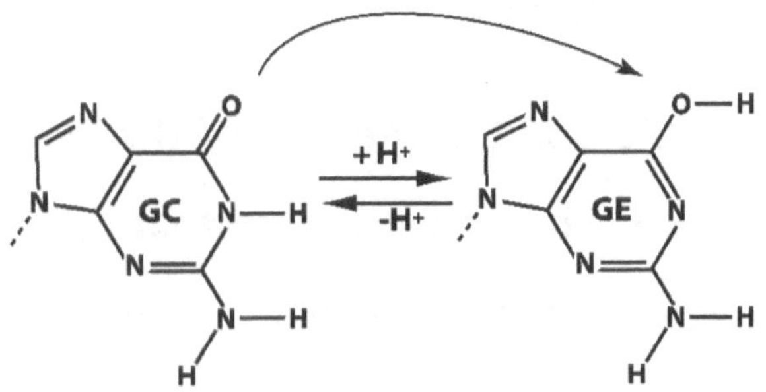

FIGURE 8
TAUTOMERISM IN GUANINE: IF THE MEDIUM IS ACIDIC THE KETONE WILL CHANGE TO AN ALCOHOL

In a prospective summary, out of the 5 bases that make up DNA and messenger RNA, we will be left with only the guanine G base and the uracil U base with the most feasible ketone spatial configurations so that they can coexist in equilibrium with their enol when the cellular environment becomes more acidic, as can be seen in figure 8 for the guanine base.

After having carried out an analysis from the electronic point of view, or the chemical investigations, we have concluded that our research into the origin of cancer will be based on these two evidences: firstly, on the phenomenon of tautomerism, which is evident or more likely to happen to both the ketonic guanine base and the uracil base. Because the correct coupling or not between these two bases depends on the modification that the chromosomes must make in order to change the electronic structure of normal DNA. Since, in normal DNA, the bases are joined together by hydrogen bonds (dotted line in Figure 8, H---O=C<, H---N<).

Secondly, and as we will see in more detail in the case of methylation, this electronic change happens, because the consumption of another animal's meat brings a jumble of cells and proteins. In addition, the meat brings the cholesterol that is specific to each animal species; therefore, the consumption of cholesterol from meat causes heart attacks, because the cholesterol of animal meat is similar to the cholesterol of a human being; therefore, the cholesterol of animal origin is insoluble in the blood of a human being. Cholesterol consumed from meat forms a plaque that adheres to the inner epithelial membrane of the arteries of a human being, causing a narrowing of the arteries; this is known as Bernoulli's principle. A blockage in the arteries causes high blood pressure, and high blood pressure can lead to a heart attack.

An example of why human cholesterol is different from animal cholesterol is because the cholesterol molecule has eight chiral centres, so that if you place a cholesterol molecule in front of a mirror, you will see 2^8, 256 different forms of cholesterols or different mirror images.

From cholesterol, sex hormones are derived; so that, from human cholesterol, children are begotten, and from pig cholesterol, piglets are begotten; from chicken cholesterol, chicks are begotten; that is to say, without cholesterol, physical life would not exist.

Proteins, because they contain an abundance of the amino acid methionine, methionine will cause methylation to occur and cancer will be induced, since the amino acid methionine, on losing its methyl group, will become the amino acid homocysteine; which, in addition to leaving us the methyl group in abundance so that the bases cytosine and uracil are both converted into thymine, it turns out that this amino acid homocysteine, which is generated inside the cells, is also an antioxidant agent; therefore, the amino acid homocysteine will usurp the antioxidant role of the natural antioxidants that are inside the cells, such as: the enzyme superoxide dismutase, alkaline phosphatase, hexokinase and oxidised NAD^+ and reduced NADH; which, as we have seen, have the function of changing the oxidation state of the iron in haemo-globin, so that haemoglobin alternately transports oxygen and carbonic acid.

If the fluid inside the cells is more acidic than normal, of course cellular respiration will stop, because the Warburg ferment will be affected, and the cells will start to produce energy in the same way as they did when we were a spermatozoon. That is, the energy at that time was produced without the need for oxygen. There is a difference, because when we were a spermatozoon, energy was produced by the fermentation of fructose, and fructose fermentation does not produce lactic acid. However, when we are already a human being engendered in the womb, energy will be produced by oxygen and glucose; but the supply of glucose and oxygen will depend on our mother's breathing and feeding.

Then, at birth, our parents will teach us to kill animals for food, because they claim that we need to consume animal proteins in order to live. We do not need proteins to live, but the amino acids that these protein chains contain, which we can find in a more abundant and varied way in the different and varied vegetables.

In fact, vegetarian animals, for example, hippopotamuses, gorillas, cows, giraffes, and elephants only eat vegetables to obtain their daily ration of amino acids. Humans do not need to kill other creatures to eat them, as food is most abundantly found in vegetables, but neither do we have to run after an animal to kill it. Vegetables make up 70% of the proteins that occur in nature; for example, soy, oats, and rice contain a large amount of protein; only these proteins are not complete because they do not contain all the amino acids essential to the diet. These grains are easier to digest, so we need to eat a variety of vegetables, but not a lot of one vegetable. For example, eating beans with rice will provide part of our amino acid ration for the day. Soy contains more complete protein than meat, and oats contain more protein than rice. A cow, for example, eats here and there, gathering her daily ration of amino acids. The domestication of animals, in the misnamed animal agriculture, is nothing but a deception towards our animal brothers, because of this ignorance of how the electronic body of human beings works chemically, we kill animals for food.

Acidosis makes respiration difficult in the mitochondria, so that the mitochondria will produce caloric energy by glycolysis of glucose, i.e., without oxygen. But, with the process of glycolysis, lactic acid will be generated inside the cells, and this lactic acid will worsen the situation of high acidity inside our cells.

On the outside of the cells, there are lactic acid bacteria, which convert any kind of sugar into lactic acid. So, when we grow up and consume too much sucrose or cooking sugar, the lactic acid bacteria in the mouth and in the colon will produce lactic acid, which will lead to tooth decay and colon cancer. As we said in chemistry, these bonds that form between hydrogen atoms are known as van der Waals forces.

We can repeat for clarification that, as can be seen in Figure 7, in normal DNA, the ketonic guanine base can form hydrogen bonds with the hydrogen that is bonded to nitrogen atom 1, and with the hydrogen of the nitrogen of the amino group that is bonded to carbon number 2, whereas the ketone group of the ketonic guanine base on carbon number 6 can accept a hydrogen bond. Of the three pyrimidinic bases: thymine, uracil and cytosine that are in the cell nucleus and can fulfil the condition of coupling with the ketonic guanine base, it is the cytosine base. There is no other pyrimidine base that has the same electronic characteristics as the cytosine base for binding to the ketonic guanine base. But, in addition, this bonding is achieved by both bases in a conjugated form. As can be seen in Figure 7, which shows that the ketonic guanine base contributes the hydrogen bond through the amino group attached to carbon number 2, and additionally through the hydrogen atom attached to its nitrogen number 1, while the cytosine base contributes to the formation of the hydrogen bond, also from its amino group attached to carbon number 4.

That is, this triple bond strength is reciprocal, and therefore this is the most stable form of coupling that forms normal DNA. Whereas this chemical condition cannot be fulfilled by the uracil base; therefore, the uracil base cannot bind to the ketonic guanine base or the adenine base to form normal DNA. We conclude that, normally, in DNA, the ketonic guanine base can only form hydrogen bridge couplings with the cytosine base, as there is no other base that has the chemical properties to make such a coupling.

But the case of adenine is equally characteristic, because the adenine base has only two possibilities, since the adenine base has a single hydrogen in its amino group attached to its carbon number 6. Thus, in order for the adenine base to be hydrogen-bonded to an oxygen of a base with a ketone carbonyl group, this coupling can only be achieved if the adenine base accepts a hydrogen bond at its number 1 nitrogen, so that two hydrogen bridges can be formed in this way. This chemical condition is only possible between the adenine base and the thymine base. In such a case, and as we can see in Figure 7, the adenine base could couple with the ketonic uracil base; but this is only in a relative way, because the thymine base is more basic than the uracil base. The thymine base, as we said, carries on carbon number 5 of its ring the methyl group that replaced the alpha hydrogen, so that methyl gives the thymine base greater energetic stability. Logically from an electronic point of view, the uracil base will not be able to couple with the adenine base. But there is no other base in the cell nucleus that performs with the same or similar electronic characteristics as the thymine base, or that can fulfil this condition to replace it; therefore, it is a living design, which, in addition to being surprising, is truly unique.

So, the uracil base does not fit with the adenine base or the ketonic guanine base to form hydrogen bonds in normal DNA, as long as the acidic condition within the nucleus is normal for DNA to replicate in that specific way. If the coupling between the bases did not happen in this way, the two ketone groups of the thymine base would face each other in one row of the DNA sidechain, and these two ketone groups would repel each other, breaking the sequence on that side of the helix in the normal DNA strand.

So, neither the uracil base nor the thymine base can pair with the ketonic guanine base to form a normal DNA strand. Whereas this chemical structure for normal coupling can only be fulfilled by the cytosine base with the ketonic guanine base.

As for this descriptive analysis, it seems a bit complicated to understand, but, as we said, cancer is a chemical phenomenon, so we need to know the logic of these couplings to know how cancer is produced chemically, since the DNA that gives structure to each cell is made of electronic matter, which will make the necessary adjustments between the electronic bonds, because the cells are not aware of its existence or its functioning. Although they make up living things, cells are made up of electronic matter, but the electronic matter is not aware of itself or its existence. Elsewhere in the Universe we do not have physical bodies; so, this chemical complication of the body only happens on Earth.

What we are analysing we have studied carefully, in order to find an explanation that makes logical sense, so that the reasoning leads us to the causes or origin of the cancer. But, in addition, the form of these couplings between the bases is electronic matter that was formed from electronic energy. So, this, and every form of matter, can be expected to change constantly, because it can form an infinite number of kinds of combinations between the infinite kinds of magnetic mass with the different kinds of bodily electronic matter.

In short, chemistry, as the science that studies how these electronic substances are transformed, has complex but logical characteristics. Therefore, it requires a knowledge and analytical acuity that is unique to each scientist, because that is the innate quality of the magnetic memory of each spirit, which is the energy that drives the electronic matter of a scientist's body. The desire to inquire in order to learn is part of the magnetic mass of the spirit that wishes to evolve; so, try to return again and again to reasoning, for this is really for you your way of learning and you will see that everything fits together in a right way within logic.

In a logical way the Universe originated; and, from the motion of an almatrino only two kinds of energy were formed: magnetic energy which, when integrated by the energy of memory, became conscious of itself; and, from this integration spirits were formed; and, electronic energy which was integrated by the force of gluons, and all kinds of inanimate electronic matter of the Universe was formed; but, electronic matter is not conscious of its existence.

The magnetic mass of the spirit is the conscious part of the human being. Ignorance is what drives human beings to make their mistakes; but there are human beings who are not aware of their existence, let alone how the Universe originated.

The magnetic mass, conscious or not of its existence, is what gives life and directs each one of the kinds of electronic matter that forms physical bodies; let's say: a spore, a butterfly, a worm, a hummingbird, a cow, a whale, a sheep, a horse, a human being... So, without the magnetic mass, the electronic matter that forms physical bodies would only be a kind of inanimate electronic matter; since the body would not have the energetic force of the magnetic mass to direct it. Magnetic memory is the energy that makes a spirit aware of itself, and drives it, so that the spirit and the physical body have the activity of life.

When the magnetic mass is separated from the electronic energy of the physical body, the electronic matter of the physical body will be left without the magnetic mass that gave life to the physical body. The magnetic mass will remain alive in a conscious manner, and the electronic matter of the body will remain inanimate without the energy of the spirit; but the electronic matter of the body without the magnetic mass of the spirit will continue to change over time; for the electronic matter of the body disintegrates. The magnetic mass of the spirit will continue to exist eternally, because the magnetic mass cannot disintegrate. But the magnetic mass of the spirit without the electronic matter of the body will have gained another apprenticeship in this new opportunity, or during the time it was part of a physical body.

The process of anchoring the magnetic mass with the electronic matter of the body without memory begins with the sexual act between the sperm in the gonads of all male animals and the ovum in the female being in order to form a functional living being. The functionality of a living being is due to magnetic memory. Spermatids have no magnetic memory because spermatids are made of electronic matter. However, the spirit with its inherent magnetic memory will be incorporated into its new physical body 5 months after gestation, when the baby's physical body is clearly defined as being the physical body of a boy or a girl.

Spirits have gender, what spirits do not have is sex. Therefore, a spirit cannot beget another spirit. On Earth there is a cycle of life; it is a cycle that is formed in the coming and going from the spirit world to the physical world, and from the physical world to the spirit world.

Presumably, such anchorage occurs first in the chromosomes, and then separation occurs again when the magnetic mass of the spirit is forced to leave the electronic matter of the body. As we see, it is perhaps not easy to be born as a baby, but it is this difficulty that makes us reflect on whether we will be born again. Unless we have an important mission to accomplish on Earth. As the electronic matter form of the body is changeable, it will undergo the deterioration of ageing. So, the changing quality of electronic matter cannot be slowed down; or it cannot remain static. The Universe and bodies have to be in motion, so that more energy is produced, and more energy produces more motion.

The evolved human being should be at the forefront in caring for and guiding those beings of other races, who likewise have to evolve each in their own way with their own form of learning.

Learning, in turn, has a scale in each race, but it is something that is its own or that is gained in its own way and at every opportunity by each individual being, even if he or she is in the same race. Electronic matter without magnetic mass cannot form a living being. The magnetic mass of any living being does not contain electronic matter; therefore, the magnetic mass of the spirit cannot merge with the electronic matter of the physical body. But it is equally impossible for a spirit to merge with another spirit because there is no such possibility since the magnetic mass of the spirit is already integrated as magnetic energy. The spirit will remain a magnetic mass that is particular and eternal.

So, given this phenomenon of coupling, this is the result of the combination of these two kinds of energy by a condition that we now say is chemical in character, which is vital for physical life to manifest through the correct coupling of the bases in normal DNA. For the electronic matter forms a code that gives the physical characteristics to each physical individual.

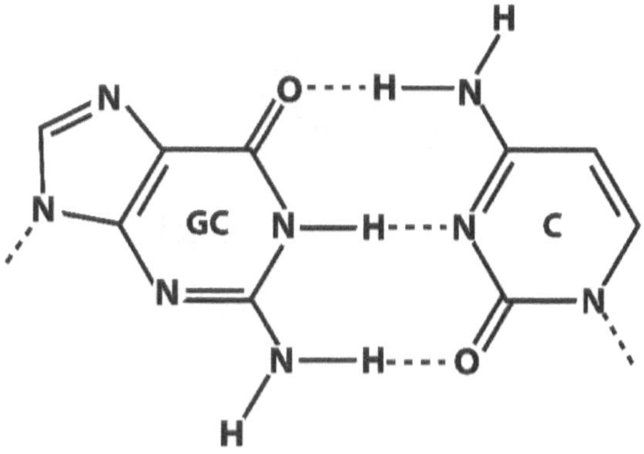

FIGURE 9
CYTOSINE-COUPLED GUANINE AND THYMINE-COUPLED ADENINE HYDROGEN BRIDGING IN NORMAL DNA

For the integration of the two kinds of energies to have that functionality or life form, the purine bases can be coupled with the pyrimidine bases in an unrestricted way, or only in that way: the ketonic guanine base coupled with the cytosine base, and the adenine base will bind only with the thymine base. As we said, the uracil base does not meet these conditions; therefore, the uracil base cannot participate to form part of DNA, as shown in Figure 8.

Chapter 6
DNA BASE LINKAGE ERROR

In normal DNA, the four bases will be hydrogen-bonded, and they will be paired in DNA in the way we have already mentioned: the adenine-thymine base pair, and the ketonic guanine $\equiv$ cytosine base pair, joined by two and three hydrogen bonds respectively. Another hydrogen bridge forms between these two base pairs: the thymine-cytosine pair. It is this thymine-cytosine coupling between the two base pairs, adenine=thymine and ketonic guanine $\equiv$ cytosine, that causes the DNA molecule to twist to the right into a helix. The base pairs bind to a chain of nucleotides that form the final DNA template, giving it stability and reliability so that the DNA molecule at normal acidity can replicate without error.

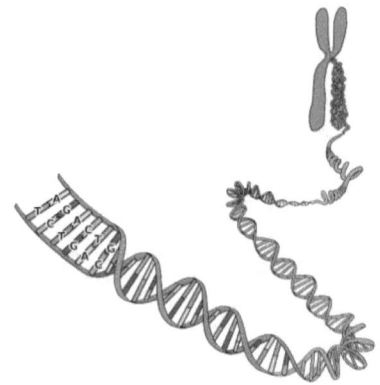

FIGURE 10
A DNA MOLECULE, THE MOST SPECTACULAR MOLECULE IN CHEMISTRY;
BECAUSE DNA IS THE MOLECULE THAT GIVES LIFE TO ALL BEINGS THAT EXIST
ON EARTH

The linkage in this stretch of DNA, as we can see in Figure 10, is more complex than the simple linkage between the bases adenine=thymine (A=T) and ketonic guanine $\equiv$ cytosine (G $\equiv$ C); and, the thymine-cytosine (T-C) pair makes a crowded, right-turned elliptical chain of the form: (A---T)-(T---C)-(G---C). It is the most fascinating molecule known to chemistry because DNA is a molecule that is active and self-replicating; but it is also the molecule that gives biological activity to all forms of physical life, so that the magnetic mass of the spirit of any living thing on earth is anchored. The DNA molecule is the one that has that functionality of offering itself as a shelter for the spirit, which is made by magnetic mass, to couple with the electronic matter of the physical body.

However, it will not be possible to merge the magnetic mass of the spirit with the electronic matter that forms the DNA; for, the magnetic mass of the spirit does not contain electronic matter; therefore, it will only be a temporary coupling between these two forms. It is a temporary cooperative effect between the magnetic mass of the spirit and the electronic matter of the body. As mentioned, it is the magnetic mass that triggers the manifestation of the life of the electronic matter of a physical body, the manoeuvring of which is being consciously or unconsciously directed by the magnetic mass of the spirit.

That is to say, the hydrogen bonds of electronic energy are what provide the strength and stability of the bond, so that the bases are held together, stable, reliable, and compact between the two chains that make up the DNA macromolecule, and the various messenger RNAs that are derived from this DNA.

It is thanks to these temporary couplings between the hydrogen atoms with the oxygen atoms and the nitrogen atoms that the chemical energy that sustains and facilitates life in all the forms and infinite number of combinations that we can imagine is configured. For example, by combining the 2 binary numbers 0 and 1 we can write all the e-mails that are transmitted in text form on Earth, but DNA forms an electronic text with 4 letters A-T-T-C-G-T; so, the combinatorics that the DNA molecule makes through genes is infinite. Therefore, there are a very large number of living beings formed by the same kinds of bases; let's say, caterpillars, ants, cows, bulls, cats, spermatozoa, and an infinite number of human beings.

At the lateral ends of the DNA molecule, nucleotide bonds have been formed by ligands with the sugar molecules deoxyribose and phosphoric acid. The DNA molecule is a spiral-shaped helix, which is twisted from left to right; since the deoxyribose molecules are right-handed, they are arranged in a sequential order. Therefore, a left-handed deoxyribose sugar cannot intervene to form DNA, because a left-handed molecule would not fit on the right-handed strand. This form of coupling between left-handed and right-handed molecules is due to chirality. The step bonds are due to both chirality and the strength of the hydrogen bonds that occur between pairs of the purine bases adenine and ketonic guanine with the pyrimidine bases thymine and cytosine. Whereas, the continuous side lines, which link these base pairs, are formed by the coupling of the sugar deoxyribose and phosphoric acid.

As already mentioned, tautomerism can only happen to the ketonic guanine and uracil bases when the chemical environment of the cell nucleus becomes more acidic. As we can see in Figure 7, when tautomerism occurs, the ketone group on carbon number 6 of the ketone guanine base or number 4 of the ketonic uracil base will convert the ketonic guanine and uracil bases into enolic bases. That is, guanine and uracil in ketonic form became a form of bases that, instead of giving, now accept electronic charges in order to form hydrogen bonds. The guanine

and uracil bases were ketonic, and that electronic configuration made them nucleophilic bases; that is, they were
Lewis's bases. But logically, when the acidity is high in the
cell nucleus, the guanine and uracil bases become enolic
bases, i.e., they will now behave as electrophilic bases or
Lewis's acids.

Whereas the alpha hydrogens, i.e., the number 1 of
guanine and the number 5 and 3 of uracil in Figure 7, being weak bonds, these hydrogens may be prone to leave
easily when a change of acidity to a higher value occurs.
Thus, the hydrogen of the amino group on carbon number 2 of the guanine base ring will remain an acceptor of
electronic charges. However, when the uracil base is converted to an enolic base, it loses the hydrogen on nitrogen 3, and so cannot form a hydrogen bridge at that site.

The amino group on the guanine will continue to form
the hydrogen bridge, as we can see in the case of the
enolic guanine. Therefore, the nitrogen number 1 of the
guanine base is left without hydrogen, which is why the
guanine base in its enolic form cannot form a hydrogen

bridge at this site, i.e., there is no alpha hydrogen in the guanine base in its enolic form, which can be released to form a double bond in the ring and give the enolic guanine base aromatic stability. If there is no aromatic stability, DNA with the enolic guanine will replicate faster, as in the case of cancer.

The guanine base with its enolic form will be able to form a hydrogen bridge with the nitrogen atom that was left without the alpha hydrogen. But the only base that can provide the hydrogen to form such a hydrogen bond is the thymine base. As can be seen in Figure 7, due to tautomerism, the uracil base became similar to the cytosine base, or the uracil base does not have a hydrogen atom on nitrogen number 3.

But there are no longer cytosine and uracil bases in the nucleus, but the thymine base with its ketonic form as soon as the guanine base and the uracil base become bases with an enolic electronic configuration. So, to form a coupling with the guanine base in its enolic form, the

only base left in the nucleus of the cells for the chromosomes to form the hydrogen bond is the thymine base.

If we look again at Figure 7, this pairing requirement
can be fulfilled by the ketonic thymine base with the
enolic guanine base, because in this case of a higher acidity, the ketonic group on carbon number 4 of the thymine
base must be more stabilised. Because the thymine base
has a methyl group on carbon number 5 of its ring and no
alpha hydrogen, the thymine base is resistant to tautomerism. But it is thanks to the stability that the methyl
group gives the thymine base on the number 5 carbon.

So, uracil is lost from the cell nucleus, as is cytosine,
because these two bases will become the thymine base
when the cytosine and uracil bases undergo methylation.
Ultimately, it will be the thymine base in DNA that can
make up for the lack of cytosine and uracil because it is
the only pyrimidine base left in the cell nucleus that can
couple with the enolic guanine base. As can be seen in
Figure 11.

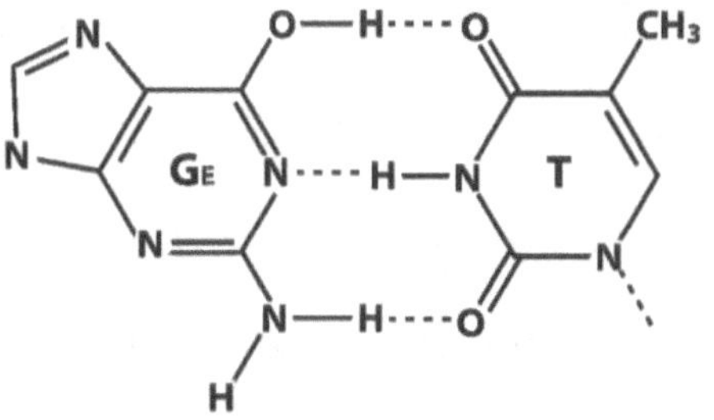

FIGURE 11
IN THE WRONG DNA, THE GUANINE BASE IN ENOLIC FORM CAN ONLY BE
COUPLED WITH THE THYMINE BASE

Regarding tautomerism in the uracil and guanine bases, look at Figure 12 to see what happens when the guanine base is transformed from its ketonic form to the enolic spatial configuration in the cell's DNA, which is a relatively more stable electronic condition under high-acidity conditions. Now, however, conditions have arisen so that instead of being with the cytosine base, the coupling of the guanine base to its enolic form occurs with the thymine base. This increased acidity, as we have said, originated from the acidic condition of the cytoplasm and then in the nucleus, which in turn was caused by excess enolic uric acid, carbonic acid, and lactic acid as a product of haemolysis and glycolysis in the mitochondria of the muscle cells, since the process of respiration was affected by the high acidity, which decreased the supply of oxygen by the normal route of respiration. This affected the oxidation/anti-oxidation system within the cell, and so on, thereby distorting the enzyme complex which before acidosis was controlled by the cell itself.

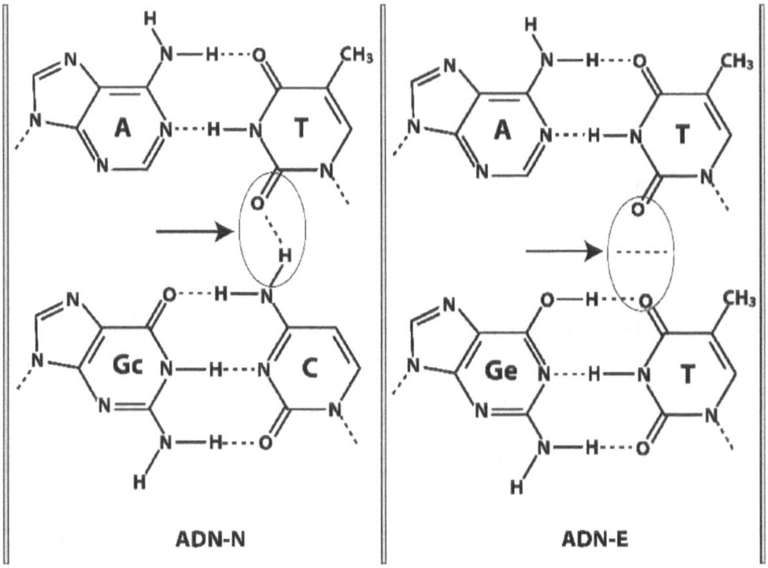

ADN-N   ADN-E

FIGURE 12
A: NORMAL DNA DNA-N KETONIC GUANINE COUPLED WITH CYTOSINE.
B: ERRONEOUS DNA DNA-E THYMINE-COUPLED ENOL GUANINE. THIS IS THE MUTATION THAT GIVES RISE TO CANCER

This whole adverse condition started, as we have shown, because of the imbalance of concentrations between enolic uric acid and sodium urate, from the moment we started ingesting the dead cells of animal meat.

$$[\text{ketonic uric acid}] \leftrightarrow [\text{enolic uric acid}] \leftrightarrow [\text{sodium urate}] \leftrightarrow [\text{H}^+ \text{protons}]$$

As we saw, we need the concentration of our antioxidant sodium urate to be at least 40 times higher than the concentration of uric acid ketone. Likewise, in the blood of a person with a terminal cancer situation, the concentration of sodium urate is 0.5 times higher than the concentration of uric acid, which must be as enolic uric acid. This means that a person with terminal cancer will not have the antioxidant sodium urate in his blood, but only enolic uric acid. So, in people with terminal cancer, there will be no antioxidants in the form of sodium urate to control oxidative stress on healthy red blood cells. But this explains why people who stop eating meat and eat only fruit as a source of antioxidants have their cancer chemically reversed, because the antioxidants in fruit reverse the acidosis and the keto-enzyme balance returns to normal. In other words, these people were relieved of cancer, because they were able to reverse the tautomerism of the guanine base in the DNA of their cells in time, and the methylation of the cytosine and uracil bases was stopped.

As shown in Figure 12, there is a probability that the guanine base with its enolic form, instead of being with the cytosine base, couples with the thymine base. Because when cytosine and uracil are converted to thymine, an abundance of the thymine base will remain in the cell nucleus after methylation occurs. This is what induces a genetic error in the DNA, because the thymine-cytokine bond is broken, resulting in a flattening of the DNA from the helix form, because the DNA is not twisted in that stretch.

As can be seen in Figure 12 on the right, the two ketone groups of the thymine base are facing each other; so, these two molecules repel each other, because no other bridge has formed in the sequence of the bases, like the one formed on the left-hand side by bridge number 3 in normal DNA. That is, where there is a bridge between thymine and cytosine (T-C). This repulsion that now occurs between the two thymine bases is what produces a break in the sequence of the DNA on the right, compared to the figure of the normal DNA on the left in Figure 12.

Cells carrying the errant DNA lose their electronic configuration, as well as their original chemical properties, and problems related to the distorted gene sequence can occur. Replication of these mutant cells induces, for example, a slight tumour, which, as it progresses in size, will become visible as a cancer as the replication of these biologically active cells proceeds. However, despite being active, these mutated cells are different from healthy cells, because mutated cells reproduce faster than normal cells. They are changing cells, because that is the nature of the electronic matter that forms the DNA to seek its electronic readjustment, depending on the acidic conditions for the chromosomes within the cell nucleus.

By provoking acidosis by the way, we feed, we have also managed to change the ketonic or normal molecular structure of guanine and uracil; therefore, the conditions necessary for the natural formation of hydrogen bonds

(H---O=C<, H---N<) will also be modified. Because in any case, the tautomeric or enolic structure of the guanine can only couple with the ketonic or normal structure of the thymine base, thus introducing a coupling error in the mutated DNA.

The original triple bond that the ketonic guanine base has to form with the cytosine base must possess that particular characteristic in normal DNA, namely that of contributing its fifth hydrogen bond between the thymine and cytosine base pairs, which, as mentioned, confers greater energetic and aromatic stability and reliability, which physically strengthens and stabilises the structure of the original DNA. Therefore, this influence as a triple bond has to be important.

As can be seen on the right-hand side of Figure 12, hydrogen bridge number 3 disappears when the guanine base in enolic form is coupled to the thymine base. That is, there is no longer a hydrogen bridge between the base pairs, as shown by the dashed line in Figure 12. Therefore, the strength of the triple bond is lower in the wrong DNA, and in a way, the wrong DNA becomes weaker, which is energetically true, because if we count them numerically, in normal DNA there are six hydrogen bonds: two (A-T), one (T-C) and three (G-C). Whereas in the wrong DNA there are only 5 bridges: two (A-T) and three (G-T). Thus, less energy will be needed to synthesise the mismatched DNA than the normal DNA, and the mutated DNA will replicate faster than the normal DNA, as in the case of cancer.

It is a transition-type mutation because it is caused by substitution between bases of the same class, i.e., pyrimidine for pyrimidine, (the base cytosine for the base thymine) which is more likely, as this form of coupling does not introduce an appreciable change in the normal molecular chemical structure, or that of the original DNA, as can be seen in Figure 12.

However, the chromosomes of a cell that are involved in this tautomerism, and if the tautomerism becomes peremptory, the cell will be able to continue biologically with its reproductive work, but with a logic of chemical character of its chromosomes. The synthesis of the DNA will be in a wrong way with respect to the other cells, or at least in terms of the speed of its replication and functionality; therefore, the mutated cell will not be able to configure the electronic matter of the body of a human being who was born with his normal DNA. However, the carnivorous human being managed to introduce a change in the structure of its genes. Therefore, these mutant cells belonging to the same body will come into conflict with the healthy cells.

The situation of higher acidity is not favourable for the coupling of the guanine ketone bases with cytosine, as soon as the cytosine and uracil bases get a methyl group that will convert them into thymine.

It is important to know, as we have said, that these differences are relative to each other, because in the electronic bonds, there does not necessarily have to be a marked contrast for the necessary adjustments to occur and for the couplings between the bases to be propitious.

In a relative sense, it can be said that if there were to be an abundance of methyl groups within the nucleus of the cells, the cytosine base would no longer be available; because in the methylation process, as we shall see, the entire cytosine base will be converted to the thymine base, which is the partner of the adenine base.

The cell nucleus, when involved in a tautomerism and methylation process, will energetically transform itself into a relatively stable and functional chemical unit under these conditions of higher acidity in the cell nucleus, so that the chromosomes replicate a DNA in the wrong way. The replication rate of the mutated DNA, although logical from a chemical point of view, will be altered from a biological perspective, and that is what shows up in what we call a mutation, as it is no longer the same molecule of the original DNA that developed in the same body made of electronic matter and magnetic mass. However, this is not a condition that can be inherited by genetic modification in all cells, because such a change in the already formed genes would be extremely complicated to happen in the same body. A person with end-stage cancer cannot give birth to a mutant, so that the newborn will carry the mutation with it, but the DNA of the newborn will replicate faster and age very quickly. A woman with high acidity, or who is pregnant and has acquired her pregnancy during the formation of mutant cells, can transmit altered DNA to the foetus, so that the child may be born with cancer inherited from the mother. We would conclude that cancer cannot be reversed in a child born with distorted DNA, but we know that cancer can be reversed in a person born without cancer.

The physical distortion of DNA can be seen in the children born after the nuclear bombs were dropped on the civilian populations of Hiroshima and Nagasaki.

What has happened is an error caused by acidosis, which alters the coupling between the bases that make up DNA, so it is possible to restore it chemically, because healthy cells are developing in a design pattern, which is determined by the traits of their genes. But it is different if we are born with a DNA that has one or several altered genes, or which already have a modified or implicit DNA structure; because this modification only has to be provided by the spermatozoon with half of its chromosomes, and the other half of the chromosomes that come from the ovum. For this mutation to be inherited, it is necessary that one of the two pairs of chromosomes is already modified. In other words, if the cancer were inherited, the genetic error could come from the pregnant mother or father.

The configuration of the phosphate groups can be changed; and the complex formed by the reducing enzymes, whose main representatives are glutathione-SH, hexokinase, catalase, superoxide dismutase, vitamin C, etc., which, as we mentioned, were the enzymes that protected the DNA against changes in relative acidity inside the cell. In other words, the right chemical and energetic circumstances are provided for the bonds to form between the enolic guanine-thymine base pairs instead of being ketonic guanine-cytosine, and so cancer originates in the physical body.

But all this marvel or ingenuity is thanks to Nature because these couplings must have happened for some very specific chemical and electronic reason. It could be, for example, the increased speed at which each different organism needs to read its codes in order to synthesise, for example, at a faster rate a certain protein by its ribosomes. Or a higher frequency of replication of their DNA in their chromosomes. Thus, each living being will have a lifespan, which will depend on the speed with which its cells replicate. This speed of replication will influence the lifespan because it is what determines the culmination of ageing, i.e., the end of the division of electronic matter in each race of living beings.

An example of this altered rate of cell replication in humans is that there is a condition called progeria, in which a child becomes old. But the speed of protein production in ribosomes has to be different from the speed of replication required in chromosomes for DNA, because if it weren't for this delay in our DNA, then we would grow old too soon; or cells would replicate before proteins are formed. A mouse, for example, can live 1.5 years; a tortoise 150 years and a human being 100 years. A whale's body is very large and can live 80 years.

Therefore, we can think that, for the first humans who did not eat meat but vegetables and fruits, uracil was present only in the messenger RNA, in order to speed up the synthesis of proteins in the ribosomes. But uracil was not in the DNA, because if it had been, DNA replication in the chromosomes would have happened more quickly. Likewise, if the thymine base were in the messenger RNA, protein synthesis would have happened too slowly.

In April 1997, an article appeared in the Proceedings of the National Academy of Sciences of the United States of America PNAS (PNAS April 1, 1997, vol. 94no. 73290-3295) by researchers Benjamin C. Blount et al. entitled: 'Folate deficiency causes incorrect incorporation of uracil into human DNA and chromosome breakage, with implications for cancer and neural damage'. This is what we called open spines in children when we mentioned folic acid. The spine is also known as the vertebral spine because the cervical vertebrae generally have a bifidus in the shape of a 'Y'. Perhaps the most important thing about this article in this case is that these researchers succeeded in demonstrating experimentally that the uracil base, which should only be in the various messenger RNAs, was mistakenly introduced into the DNA. But this thymine and enolic uracil bases are virtually identical, so we will not know whether it is the thymine base that actually causes the DNA breakage in a person with cancer when coupled with the guanine base in its enolic form.

Chapter 7
METHYLATION OF CYTOSINE AND URACIL BASES

Methylation is necessary to introduce the methyl group (-CH₃) into amino acid molecules. But, as with sodium urate from ingested meat cells, consumption of animal protein will bring in excess of the amino acid methionine; and, in this case, the methyl group of the amino acid methionine is released, leaving the methyl radical free. The methyl radical is a nucleophile, whose negative charge should be consumed on the inside of the cell by the antioxidant system, and on the outside of the cell by sodium urate. However, if the cell nucleus and blood become acidic, the methyl group cannot be neutralized, and on the inside of the cell, the methyl radical will react with the cytosine base and uracil in an enolic form to cause these bases to become methylated. As we can see in Figures 13 and 14 respectively.

In the case of the cytosine base in Figure 13, when the cytosine base captures the methyl radical, the cytosine base will be transformed into the thymine base. The same happens to the uracil base when the uracil base is in the enolic form as a result of high acidity. As shown in Figure 14, the enolic uracil base will be transformed into the thymine base. That is, the uracil base will be affected by a methylation process when the acidic medium converts the uracil base from its ketone form to the enolic form. After the methylation process, the nucleus of that cell will be left without the bases cytosine and uracil, because both bases will be converted to the base thymine.

In order to replicate DNA, the chromosomes will use the thymine base, which is now abundant in the nucleus of that cell, as a substitute for the cytosine base. Similarly, to produce messenger RNA, chromosomes will use the thymine base to make up for the lack of the uracil base in the cell nucleus.

If cells from meat of animal origin were not consumed, tautomerism of the guanine and uracil bases in human blood would not occur, because the degree of acidity would not increase. Whereas the consumption of animal protein causes methylation of the cytosine and uracil bases to occur in the nucleus of human cells.

Inside the cells, as methionine is converted to homocysteine, homocysteine will usurp the function of the cells' own antioxidants. So, in this way, it will start to slow down the respiratory enzyme system of the cells, which, as we have seen, is important within the cells to control the degree of acidity, when energy is generated in the form of heat without oxygen in the mitochondria.

Oxygen-free energy is needed in cases of distress; for example, when we are frightened, we stop breathing, and cortisol causes the insulin level to drop so that more glucose is available, in case we have to make a run for it. The process of breathing without oxygen through glycolysis is most developed in birds, reptiles, insects, and diving animals, such as turtles, seals, and penguins. Diving animals have to dive into the water to search for food, but then they have to come to the surface to breathe oxygen from the air.

The anguish in the inhospitable cages for the transfer of the cows to the slaughterhouse causes the cows to generate cortisol in order to try to escape from the anguishing confinement. When the cow's meat is consumed, humans consume the cortisol from the cow's distress. The cortisol consumed from the cow's meat causes the human being's blood to suppress insulin production, which contributes to the human being's diabetes. Humans should live free on the surface of the earth, eating vegetables and fruits and breathing air to supply their cells with electronic energy.

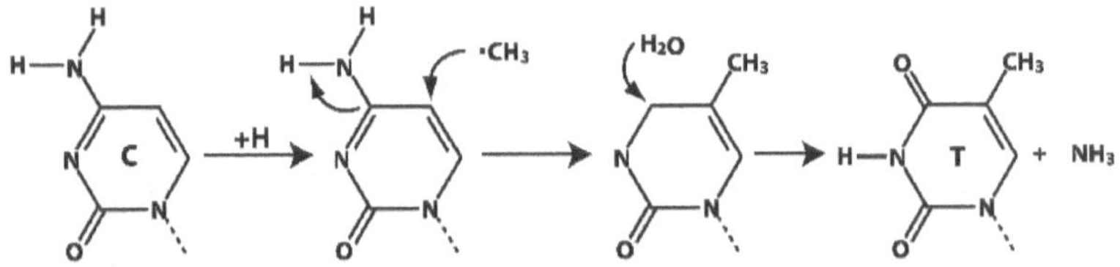

FIGURE 13
CONVERSION OF THE CYTOSINE BASE INTO THE THYMINE BASE BY METHYLATION

On the other hand, as cytosine becomes acidic, carbon 5 in the cytosine ring will become positive, i.e., electrophile, and vulnerable to attack by free radicals or nucleophiles, such as the methyl group. Being an electronic giving group, the methyl radical can react with electronic nuclei, i.e., those particles that are positively charged, as shown by the arrows in Figures 13 and 14.

In the case of Figure 13 for the methylation of the cytosine base, the methyl radical left over from the methionine will attack carbon 5 on the cytosine ring (see Figure 7 again), and this will convert it into an intermediate, i.e., 5-methylcytosine. Then, as the 5-methylcytosine compound loses the amino group at carbon 4 in the form of ammonia (NH_3), the site left by this amino group will be occupied by a water molecule. As a result, 5-methylcytosine will be transformed into the base thymine plus ammonia. When the acidity is high, the ammonia will be converted into the ammonium ion, which can be transported to the liver where it will be converted into urea by carbon dioxide. The ammonium in the form of urea will eventually be excreted in the urine. This is the source of the increased urine volume in diabetics, and the ammonia odour in elderly cancer patients.

With respect to the uracil base, this base will first be converted into its enolic form by tautomerism, and then followed by methylation. As we can see in Figures 4 and 7. Because upon conversion to the enolic form, carbon 5 of the uracil base becomes positive, i.e., uracil will be a Lewis acid. Therefore, when uracil is converted to the enolic form, the uracil base becomes more prone to attack by free radicals, such as the methyl group, which is introduced at carbon 5 of uracil, in the same way as happens to the cytosine base. The intermediate state of uracil is shown in Figure 14. So, as with the cytosine base, the methyl group will be incorporated on this carbon of the uracil enolic base; but, if the uracil base is converted to an enolic base, the uracil base will be converted to the thymine base by methylation.

In this case, just as ammonia remains as a residue in the methylation of the cytosine base, in the methylation of the enolic uracil base, a hydrogen atom ($\frac{1}{2}H_2$) must be left free, which is then converted into a hydrogen molecule H_2. This methylation reaction is possible, as we know that molecular hydrogen is a reducing agent, which is compatible with the reducing character of homocysteine within cells, but the reducing character of these two substances, homocysteine, and hydrogen, will usurp the reducing character of the cell's natural reducing enzyme system.

Perhaps most importantly, the result of the high acidity within the nucleus is that the guanine and uracil bases became enolic, and this caused the cytosine base to become the thymine base, as can be seen in Figure 13. Likewise, the uracil base from its enolic form became the thymine base, as can be seen in Figure 14.

So, this methylation process can happen by this way of demethylation of the amino acid methionine, which was incorporated into the cells in excess during the years of traditional animal protein consumption. With this ade-nine=thymine base pair there will be no problem because what there will now be is a greater amount of the base thymine.

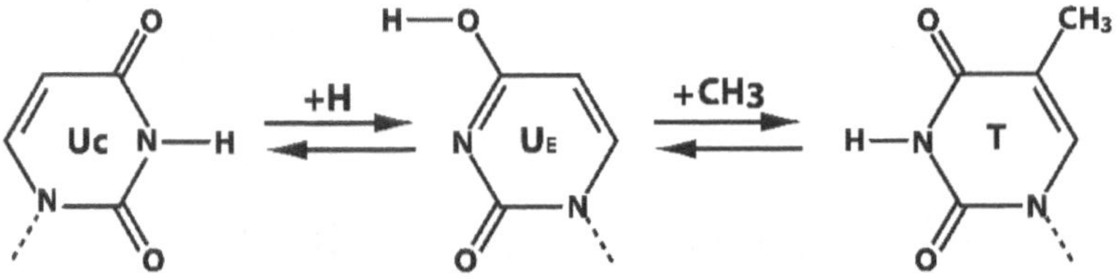

FIGURE 14
CONVERSION OF THE ENOLIC URACIL BASE TO THYMINE BY METHYLATION EFFECT

As the methylation process proceeds, the nucleus of that cell involved in replicating its DNA will at some point run out of the bases cytosine and uracil, and this would force the cell to chemically change the base pairings in the DNA by the chromosomes. As can be seen in Figure 7, when the uracil base becomes enolic, this base cannot re-place the cytosine base in the DNA; since the uracil base cannot form part of the messenger RNA because there is no longer a hydrogen on the nitrogen number 3 of the enolic uracil. The only base left in the cell nucleus for the chromosomes to make the coupling with the enolic gua-nine base and produce messenger RNA is the thymine base. Because the thymine base has a hydrogen on nitro-gen 3. But there is no other base in the nucleus of the cell with these same electronic characteristics. The only base left in the cell nucleus to form couplings is the thymine base.

Chemical conditions have arisen, which will cause a readjustment of the electronic couplings in the DNA, which will influence the function and the original struc-ture of that DNA; in other words, the cell mutates; and that nucleus will now be different, because the chromo-somes will use the thymine base as the other base for coupling with the guanine base, which is in the enolic form. It is a coupling that normally would have been oc-cupied by the cytosine base with the guanine base in its ketonic form, but not in its enolic form; and it is evident that the thymine base now participates with its abun-dance so that the chromosomes form a new DNA; but this

DNA that the chromosomes produce will be altered with respect to the normal DNA.

As said, the same will happen in messenger RNAs, since the uracil base has disappeared from the cell nucleus, and this missing uracil base will be replaced by the thymine base, which does not actually participate normally in forming messenger RNA.

So, with that excess thymine base, you can alter the transfer RNA and the messenger RNA, and with that, you can directly influence the problems related to amino acid sequencing, which will change the order of insertion of the different amino acids in the protein chains. Because as we explained, the change in a triplet induces a change in the position of an amino acid in the protein chain, and this will contribute to the exchange of one amino acid for another, but the protein chain formed will not be the same as it should have been formed.

Tautomerism and methylation can lead ribosomes to a synthesis error, as the triplets will be different. That is, the normal initiation triplets, or when there is no tautomerism and methylation, will be of the form: uracil-adenine-cytosine (U-A-C) with respect to the transfer RNA that must couple with the adenine-uracil-guanine (A-U-G) triplet of the messenger RNA. Similarly, the normal termination triplet will be uracil-adenine-adenine (U-A-A) in the messenger RNA, which has no pair in the transfer RNA; therefore, when this triplet arrives from the messenger RNA with the signal that nothing is going there, this triplet signals the ribosome to terminate the synthesis of the protein chain.

When there is no cytosine or uracil in the cell nucleus, but only thymine, these triplets carried by the messenger RNA will be different; therefore, the insertion and amino acid sequence in the protein will be different. For example, the initiation triplet will be changed to thymine-adenine-thymine (T-A-T), while the termination triplet will be thymine-adenine-adenine-adenine (T-A-A). In this way, the ribosome will not find the code that tells it where the protein synthesis will start and where it will end; thus, the protein produced by the ribosomes will be different; at the same time, the protein produced will be shorter or longer.

From that moment on, both in the nucleus and in the cytoplasm of the cell, an imbalance will be generated that will affect the structure of the entire cellular framework, and a new kind of cancer cell will replicate. But healthy or neighbouring cells that are not affected will seek electronic readjustment of the chemical structure of their electronic design and functionality. These remaining healthy cells are the cells that we must prevent in time to mutate, so that the healthy cells are not outnumbered by the mutant cells. The cancer cells will disappear as long as we act in time to prevent the healthy cells from mutating.

This will not be achieved until the healthy cells find their proper acid-base condition again. In such a case, everything will depend on the human being involved in the process of tautomerism & methylation, but it will not be the fault of our cells, because it is we as spirits made of magnetic mass who decide what we eat and what we should not eat, in order to feed our healthy cells, which are only made of changing electronic matter. In other words, the mutated cells are not aware of the genetic error that happens to them. The only thing that is aware that cancer has happened to it is the magnetic mass of the spirit, which is distressed because it has managed to change the shape of the normal framework of its physical body by means of cancer, which was caused by the bad feeding strategy of the magnetic mass of the spirit.

As far as humans are concerned, the genome is characterised by a heterogeneous genome and an arrangement of base pairs in the DNA. However, this arrangement of base pairs is not random, but depends on the electronic characteristics that are formed, which is what gives the final physical appearance to each DNA in a particular way. It is to be expected that such an arrangement or sequence between base pairs results in a truly infinite number of combinatorics. It is the base pairs that give this combinatorial possibility, although individually, the shape of these pairs in DNA has to be of the form ketonic guanine≡cytosine, cytosine-thymine, and adenine=thymine, because if the characteristics of these individual bonds are changed, this will influence the sequence of the base pairs in the final structure of each DNA.

There are abundant regions with ketonic guanine-cytosine triplet assemblies; this is possibly the result of the more stable hydrogen bond that forms between the additional thymine-cytosine base pair. As it is, hydrogen bond number 3 forms on the left-hand side of Figure 12. But the stable three-dimensional structure of DNA will be modified when the enolic guanine-thymine pair is formed, because a hydrogen bond cannot form between the thymine-thymine pair on the right of Figure 12. Unless this space is occupied by a proton from the acidic medium, but even then, this would change the normal genetic sequence of the base pairs.

What makes DNA stable is that the ketonic guanine pairs with cytosine, so that other bonds can form between the base pairs, such as the thymine-cytosine bond that twists and stacks normal DNA. The most logical way for this to happen is for the ketonic guanine-cytosine triple bonds and the thymine-cytosine pair to form between the two base pairs, which gives the normal DNA molecule greater energetic reliability. These triple pairs are the ones that contribute the most energetic power to stabilise the normal DNA. This is why the observed average content of DNA triple bonds with the ketonic guanine≡ cytosine pairs are approximately 60 % higher than the theoretically expected 50 %.

The increased diversity of triple bonds in the order of 60 % is correlated with the so-called gene richness, which means that genes have the propensity to concentrate in those regions richer in couplings with ketonic guanine≡ cytosine triple hydrogen bonds. As we can see in Figure 12, such triple hydrogen bond richness can be diminished by the effect of acid-base alterations within the cell nucleus, as in this specific case of tautomerism, which influences the generation of uracil and cytosine methylation.

On the left of Figure 12, you can see why in normal DNA there are preferential or more abundant regions in the pairs formed by the triple ketonic guanine≡cytosine hydrogen bridge. Because in this type of crowded or clustered molecules such as DNA and messenger RNA, what exists is an interaction between electronically coupled clouds; therefore, this DNA and messenger RNA are changeable, or a rearrangement is encouraged to occur; and the chemical stability of the three-dimensional structure of DNA will depend on the force of attraction with which each molecule, or groups of molecules contribute to the readjustment of the electronic matter of DNA.

Triple hydrogen bonds cause the chain-like molecule to twist into a spiral shape as each pair joins the ribonucleotide chain. So, the DNA chain is twisted to the right, which happens, as mentioned, because the deoxyribose sugar involved in the configuration of DNA is of right-handed spatial configuration. So, in the left-hand strand of Figure 12, the most likely thing that can happen is that the pair of adenine=thymine hydrogen double bonds will appear, but reversed, which will go on to make up the coding structure of that gene.

The completion of this sequence, i.e., for the various genes to form in the DNA molecule, the length of the DNA chain cannot be infinite, so these hydrogen bonding forces are weakened. Therefore, no additional base pairs will be allowed to be incorporated into the DNA chain. This is what determines the final physical pattern of each DNA.

On the right-hand side of Figure 12, we find the same situation, but in the wrong way, due to the presence of the thymine base in the base pair enolic guanine-thymine, because in this case, there is no cytosine base in the nucleus of the mutated cell.

As we can see in Figure 12, the formation of that second hydrogen bridge between the two base pairs no longer exists, because the two ketone groups of the thymine base repel each other on that side of the wrong DNA strand. Of course, the bond strength becomes weaker in this case; therefore, in enol form, the bond strengths will be weaker; and the result is that the bond strength of the ketonic guanine≡cytosine triplet is greater than that of the enolic guanine≡thymine triplet. For, although a triple hydrogen bonding bridge has formed between the enolic guanine ≡ thymine bases, this will be an energetically weaker DNA, because the thymine-cytosine hydrogen bond has not formed between the two base pairs adenine=thymine and ketonic guanine≡cytosine. This configuration without the thymine-cytosine hydrogen bond will energetically contribute less to the formation of abundant zones for that gene containing the wrong pair of enolic guanine-cytosine; because this errant DNA will be energetically easier to synthesise; but it would bring less stability with its binding strength to the DNA molecule of a person with cancer, compared to how the normal base pair of ketonic guanine-cytosine in the DNA of a normal person did with greater strength.

DNA is what gives each organism its physical blueprint; it is the original blueprint that is laid down in the nucleus of each cell. It is a genetic code; therefore, by changing the structure of the DNA, the original physical imprint of the human being with which he or she was born electronically will also be modified.

It will be a logical integration between the electronic charges of the electronic matter; because in chemistry, the final product that results will always be the most stable when the difference in electronic charges is the most intense, even if it is the most difficult to synthesise energetically; since, what counts in the end, is the electronic stability or the lowest energy contained in the resulting product.

The lower energy required to form a weaker binding force will help this mutant DNA to replicate faster, but it will ultimately be a more unstable DNA compared to normal DNA. Because the hydrogen bonding between the ketonic guanine $\equiv$ cytosine bases gives the normal DNA greater stability, compared to the case that forms with the coupling error between the enolic guanine$\equiv$thymine bases.

Once these conditions for the chromosomes to synthesise the wrong DNA have been achieved, the gene can lose both its sequentiality and its replication rate, because the life span of each being will depend on how fast it replicates. In which case, a carrier cell with such an error will be different because of the mutant factor. Thus, a sister cell originating from this mutated cell will copy the same error in the future, until a large group of mutant cells is formed. As a consequence, some cells will be replicating faster than others, resulting in the formation of a lump or outgrowth of mutant cells that will become visible in the form of a tumour. In addition to other types of genetic diseases, or diseases that influence the occurrence of these inconsistencies in the archetype inherited by the normal human individual, but which have been affected by an error of a genetic nature.

The lower energetic force needed to form the enolic guanine-thymine triple hydrogen bond will lighten the synthesis of that erroneous DNA, as we said, so that the presence of uracil in RNA, but not in DNA, may be a controlling chemical mechanism available to cells to accelerate the speed of protein production, but at the same time to slow down the speed with which DNA is replicated by chromosomes in the cell nucleus. In other words, it is this order that determines the rate at which DNA replication takes place in the chromosomes of the cell nucleus in each living being.

Perhaps it is because of this difference in energy synthesis that the lower energy invested in the formation of the mutated DNA means that in mutant cells, the speed of replication is chemically accelerated, as can be seen in the accelerated growth of cancer. This mismatch between the bases makes sense from a chemical, i.e., energetic point of view, where the influencing factor is the changing character of the electronic matter of the physical body.

Whereas the magnetic mass of the spirit cannot be altered in any way and in any sense because the magnetic mass of the spirit is a more stable silhouette than the more stable electronic matter, because the magnetic mass of the spirit contains no nuclei and no electrons. The magnetic mass of a spirit is more compacted or bound together than the more stable electronic matter.

These changes to the DNA can be tolerated as long as the number of mutant cells does not exceed the number of healthy cells, or the whole organism does not collapse completely. As it turns out, the cell body will not be able to withstand this accelerated growth of mutant cells for long, because the chemically logical rate of replication and functionality does not correspond to the same conditions as the originally formed human being.

The distorted process can be reversed, but only if the person realises that the cancer problem is chemical in nature, and this will happen if the dietary strategy can be changed in time.

In December 2007, one of the members of the British Whitehead Laboratory research group, Rudolf Jaenisch, demonstrated that there is a link between the phenomenon of methylation and the development of colon tumours in mice. For this researcher, methylation is the accumulation of excess methyl groups in certain regions of DNA. Rudolf Jaenisch was able to deduce that methylation causes the deactivation of the gene that monitors the correct functioning of DNA, or that this is the gene that has the task of repairing or reversing what could be the beginning of a genetic error, and as a consequence, incites the formation of small polyps. It was also found that methylation increases the frequency of intestinal tumours in mice by 60-100% and, on average, significantly increases the growth of microscopic tumours.

According to Rudolf Jaenisch, DNA methylation has been correlated with the development of cancerous tumours in humans, because it is a type of chemical change in DNA that can be inherited, provided that the changes are tolerable. This would also explain why cancer occurs in children who have not eaten enough meat at a young age. In this case, the methylation was inherited from the mother as explained, because the baby in the womb was fed what the mother was fed. Because it is an inherited kind of mutation, it will be more difficult to reverse, because it is part of the whole genetic makeup of the child, whose DNA works by chemical but not biological logic.

Whereas normally, in a person who was born healthy, this error could be repaired without appreciable changes in the original DNA sequence. But only the intervention of the natural chemical environment of the cell itself is necessary, so that the reversal of this caused modification can be helped by a timely return to a vegetarian lifestyle, i.e., by consuming plant foods, or foods that are suitable for the cellular structure of a human being.

The opportunity could be given for the cellular system to return to its normal acidity or pH conditions. In other words, this process of reversing methylation and tautomerism would be what causes the progression of cancer to be chemically interrupted on its own, since the cells have the mechanisms and their own action to correct these anomalies themselves, which we have induced through our own fault.

How? By avoiding the consumption of sugar in the form of sucrose, dairy products, vegetables rich in oxalic acid and fizzy drinks, as the enzyme carbonic anhydrase will convert the carbon dioxide contained in fizzy drinks into carbonic acid. Definitely refrain from consuming meat of any kind until the accelerated growth of the mutated cells can be stopped, but if you want to continue being a carnivore, the same cancerous condition will surely reappear with greater intensity.

The other factor contributing to the methylation anomaly is the consumption of cow's milk products, as diluted cow's milk protein contains up to 150% more methionine when breast milk is chosen for reference. But humans are the only ones who continue to drink milk after weaning has taken place, i.e., when they no longer need to drink cow's milk, which is actually taken away from the calf. Whereas no other living creature knows how to make cheese from its own milk. When it comes to cheese made from cow's milk, the methionine content of this product is 900 % higher than that of mother's milk, although it has not yet occurred to humans to make cheese from women's milk, which is the only cheese of animal origin that humans can eat.

So, this healthy feeding effect is what regulates, for a normal process to occur through the mechanism of genetic modification, because adaptive activity of certain genes is also necessary, in those inherited regions of the genome, depending on what the cells need to express or do at a given time. As all cells that make up the same organism possess identical DNA, the covert sequence of that DNA will be a key element for the identity to be inherited by the future cell, but in addition, a logical mutation is necessary as long as no bases in the DNA are swapped, as only the sequence of genes must be altered to produce other life forms, or a wide variety of distinctive cells found in the body of a human being.

It is also a necessary natural reason for the improvement and perfection of each race. For instance, more beautiful women and more intelligent children are born every day, and it will be the more attractive and intelligent men and women who will contribute most to the improvement of their own future race. But also, this is a game played by insects such as ants and bees, or animals who compete, and only the female and male will remain who have the greatest energetic strength; and, therefore, a high genetic capacity for the betterment of their race.

When a cell divides, it may incorporate that enhancement or upgrade, but its original quality for enhanced functionality cannot disappear, as in the case of tautom-erism and methylation. For, the natural methylation, characteristics and sequential pattern must be main-tained in the harmony and genetic memory of the new cell to be formed. For example, if the new cell that origi-nates belongs to the heart, it must copy the function of its parent in order to inherit the same instructions on how to contract and dilate in order to continue the work of ejecting blood. If the cell is modified by tautomerism and natural methylation, this memory of function is lost, and the new cell that emerges can no longer perform the function of its ancestor. The correct sequence of guanine, cytosine, thymine, and adenine bases in the cell's DNA is what allows them to replicate without error, but they must also carry the instructions that must appear in the new cell that is formed.

In cells, as mentioned, it is the ribosomes that carry out the task of protein synthesis, and similar to the exam-ple of reading a misspelled text, they have to recognise and analyse the sequence properly, in order to minimise the probability of introducing a triplet reading error, which could lead to a misleading result of confusion and function, with respect to the appropriate proteins produced by healthy cells. This process will depend on the degree of acidity within the cell, and more specifically within the cell nucleus. The Golgi apparatus is the organelle that inspects the functionality of the proteins produced in the ribosomes.

Let's say, this was a rather careful analysis, to know how our cells function chemically, and what is the kind of electronic energy of the electronic charge differences that makes them function, to give life to the electronic matter of the physical body; that is, so that the electronic matter, can be transformed into other forms of electronic matter and electronic energy, thanks to the intervention of the magnetic mass of the spirit, who is the one who directs and gives life to the electronic matter of his physical body. The spirit that in the future will be incorporated into a new physical body is obliged to know how its physical body and the physical bodies of other living beings function chemically. But perhaps, at the present time, the analysis of body chemistry is a little complex for those who do not have the tenacity to want to learn how the cells of their body function, which is actually made of electronic matter but animated by the magnetic mass of the spirit.

Our intention, with these books, is for humanity to change its way of thinking and acting; since, through ignorance of their origin, human beings are destroying themselves and the forest, and breeding animals to kill them and devour the flesh of their electronic bodies in a harmful meat industry. Maybe animals are unaware of their existence and the existence of the Universe, but what all animals do have is feelings.

We must act in time, in order to save animals and Planet Earth. If the human being suffering from cancer manages to save himself, even if he is not aware of why, or how the cancer disease happened, he would be an unconscious human being. Our motive to be born again would be to be able to live free of organic diseases; and in harmony with all living beings, because we all have the right to exist; since all living beings emanate from the same energy that forms the Universe.

THE AUTHOR'S WORK

Graduated from the School of Chemistry, Faculty of Sciences, Universidad Central de Venezuela, with a degree in Chemical Technology. Postgraduate studies in Food Science and Technology. Special work on the chemistry of natural products and the chemistry of diseases. Chemical process designer. These books should be subject to revision as we become clearer about how the Universe was formed, so try to read the latest edition of each book. These books are: "The Chemistry of Cancer". "The Chemistry of Diabetes. "The Heart Attack". "Alzheimer's". "The Chemistry of Arthritis". "The Chemistry of Thought". "The Chemistry of the Spirit". "How the Universe was formed". "The Expensalists". "Why You Shouldn't Eat Meat". "The Micro World". "Does God Really Exist?". "Objecting to Albert Einstein's Relativity". "Divining the Future". "The Mistake of the Great Scientists". "Life on the Sun". "The Universe before Zero Time". "The Energy of the Spirit". "The Origin of Cancer". "The World of Cells". "The Chemistry of Disease". "The Particle that Created the Universe". The Chemistry of Cancer, seventh edition. The Chemistry of Diabetes sixth edition; The Chemistry of Heart Attack fourth edition, "The Chemistry of Memory"; The Chemistry of Arthritis third edition. "The Creative Power of the Mind. The Particle that Formed the Universe, third edition. "The Initial Mass of the Universe". "You Shouldn't Eat Meat". "The Origin of the Body and the Spirit". "Worship the Universe". "Sugar an Enemy in the Kitchen". "Time Travel". The Chemistry of Cancer Edition 8. The Chemistry of Diabetes, Edition 7. The Chemistry of Heart Attack Edition 5. The Memory of the

Spirit Edition 1, The Chemistry of Arthritis Edition 5. "The Life of Spirit". "Rewriting Science". "The Beginning of the Universe". "Spiritual Growth". "Coupling of the Spirit with the Body". "The Origin of Life". "Death Does Not Exist". The Chemistry of Cancer Issue 8. The Particle that Created the Universe". "The Inaugural Moment of the Universe". "The Equation of the Universe". "The Infinitesimal Universe". "The Book of the Universe". I Came from the Sun, Second Edition. Death Does Not Exist, Second Edition. The Shortest Book, but which explains how the largest system in existence was formed. "The Spirit Syndrome. "The New Science. "The Almatrino. The Origin of the Universe First Edition. "God Does Not Exist. Death Does Not Exist Third Edition. The Origin of the Universe Second Edition.